미국 수학/국제학교 수학을 요리하다!
미국 현지 20년 경력 Brian 선생님의 미국 수학 만점 레시피!
✹ 미국 아마존 스테디셀러 ✹

R E C I P E

이연욱(Brian Rhee) 지음

IGCSE
Additional Mathemathics

IGCSE Additional Mathematics
실제 시험에 나올 수 있는 개념과
문제 유형을 한번에 해결한다!

HERMONHOUSE

Preface

미국 버지니아 수학학원 원장 및 강사로서 현장에서 20년 이상 학생들을 지도하는 동안 수많은 교재를 연구하고 분석하여 IGCSE Additional Mathematics 및 IB Math를 수강하는 학생들이 좋은 성적을 받도록 도움을 주었습니다. 하지만, 항상 안타깝게 생각 하는 있던 것은 유명한 출판사의 교재들이라 할지라도 실제 시험 출제 경향이나 스타일과 차이가 있는 경우들도 간혹 보이고 턱없이 부족한 개념설명과 문제 난이도로 인해서 진정으로 학생들의 수학실력 향상에 도움을 주지 못한다는 것입니다. 그럼에도 불구하고 적절한 대안이 없어서 그 교재들을 사용하여 학생들을 지도해야 한다는 답답함이 있었습니다. 마음속에 있던 이러한 안타까움은 좋은 교재를 만들어 보자는 용기로 바뀌어 집필을 시작하게 했고, 온/오프라인의 많은 지도경험을 통해 얻은 이해하기 쉬운 개념설명과 핵심포인트를 교재에 소개한다는 목표를 가지고 노력한 결과 집필을 마칠 수 있었습니다.

이 책은 방대한 IGCSE Additional Mathematics의 토픽들을 15개의 토픽으로 세분화하였습니다. IGCSE Additional Mathematics 시험에 출제될 수 있는 모든 개념과 실전시험에 나오는 모든 문제유형들을 포함하여 실전시험에 대한 적응력을 높일 수 있도록 구성했습니다. 이 책을 통해서 중학교 과정의 탄탄한 수학 실력을 쌓을 수 있는 것은 물론이고 향후 공부하게 되는 IB Math나 A-Level 등의 이후 고등학교 수학에 있어서의 든든한 기초가 될 것입니다.

이 책은 미국 및 한국의 학원교재와 1:1 개인지도 교재로 기획되어 미국에서 먼저 출판하였으며, 미국 아마존 닷컴에서는 스테디셀러로 학생들의 인기를 얻고 있습니다. 동시에 No.1 유학전문 인터넷 강의 사이트인 마스터프렙(www.masterprep.net)에서 IGCSE Additional Mathematics의 교재로 사용되고 있으며 학생들의 좋은 반응으로 수학 교육자의 한 사람으로서 보람을 느끼고 있습니다. 책의 기획 의도상 교재에는 Answer keys만 포함되어 있고 상세한 풀이는 포함되지 않았습니다. 자세한 풀이를 원하는 학생들은 지도하시는 선생님께 도움을 요청하시거나 마스터프렙의 인터넷 강의를 듣는 것을 추천합니다.

한국에서 교재를 출판할 수 있도록 도와주신 헤르몬하우스와 항상 저에게 용기와 격려를 주시는 마스터프렙의 권주근 대표님께 진심으로 감사드리고, IGCSE Additional math를 공부하는 학생들에게 도움을 많이 주는 교재로 남기를 기 대합니다.

2023년 6월
이연욱

이 책의 특징

1 방대한 IGCSE Additional Mathematics의 방대한 토픽들을 15개의 토픽으로 세분화하였고 실제 시험에 나올 수 있는 모든 개념을 설명했습니다.

2 교재 구성은
1) 개념설명
2) 예제(example)를 통한 개념이해
3) Exercise Problems를 통한 개념적용방법 극대화

3 위 3단계 방법론을 적용하여 체계적이고 효과적으로 접근할 수 있도록 했습니다.

4 실전시험에 나오는 모든 문제유형들을 포함하여 실전시험에 대한 적응력을 높였습니다.

5 2027년까지 바뀌는 syllabus를 책에 모두 반영하였습니다.

저자직강 인터넷 강의는 SAT, AP No.1 인터넷 강의 사이트인 마스터프랩 (www.masterprep.net) 에서 보실 수 있습니다.

3

저자 소개

이연욱 (Brian Rhee)

미국 New York University
미국 Columbia University
미 연방정부 노동통계청 근무

No.1 유학생 대상 인터넷 강의 사이트 마스터프렙(www.masterprep.net)
    수학영역 대표강사
미국 버지니아의 No.1 수학전문

미국 버지니아의 No.1 수학전문 학원인 솔로몬 학원(Solomon Academy)의 대표이자 소위 말하는 1타 수학강사이다. 버지니아에 위치한 명문 토마스 제퍼슨 과학고(Thomas Jefferson High School for Science & Technology) 및 버지니아 주의 유명한 사립, 공립학교의 수많은 학생을 지도하면서 명성을 쌓았고 좋은 결과로 입소문이 나 있다. 그의 수많은 제자들이 Harvard, Yale, Princeton, MIT, Columbia, Stanford와 같은 아이비리그 및 여러 명문 대학교에 입학하였을 뿐만 아니라, 중학교 수학경시대회인 MathCounts에서는 버지니아 주 대표 5명 중에 3명이 바로 선생님의 제자라는 점과, 지도한 다수의 학생들이 미국 고 교 수학경시대회인 AMC, AIME를 거쳐 USAMO에 입상한 사실들은 선생님의 지도방식과 능력을 입증하고 있다.

아마존 닷컴(www.amazon.com)에서 미국수학전문 교재의 스테디셀러 저자이기도 한 선생님은 SAT 2 Math Level 2, SAT 1 Math, SHSAT/ TJHSST Math workbook, IAAT와 AP Calculus AB & BC 등 다수의 책을 출판하였고, 지금도 여러 수학책을 집필 중이며 한국에서도 지속적으로 선생님의 책이 시리즈로 소개될 예정이다.

고등학교 때 이민을 가서 New York University에서 수학학사, Columbia University에서 통계학석사로 졸업한 후, 미 연방정부 노동통계청에 서 통계학자로 근무했으며, 미국수학과 한국수학 모두에 정통한 강 점을 가지고 있을 뿐만 아니라 Pre-Algebra에서부터 AP Calculus, Multivariable Calculus, Linear Algebra까지 지도할 수 있는 실력자이다. 또한 미국 수학경시대회인 MathCounts, AMC 8/10/12와 AIME까지의 고급 수학을 모두 영어와 한국어로 강의를 하는 20년 경력의 진정한 미국수학 전문가로 탄탄한 이론은 물론, 경험과 실력과 검증된 결과를 모 두 갖춘 보기 드문 선생님이다.

CONTENTS

CHAPTER 1   Functions   9
  1.1 Definition of a function   9
  1.2 Composition functions   13
  1.3 Modulus or absolute value functions   14
  1.4 Inverse functions   16
  1.5 Transformations   19

CHAPTER 2   Quadratics   29
  2.1 Quadratic functions   29
  2.2 Maximum and minimum value by completing the square   31
  2.3 Solving quadratic equations   32
  2.4 Solving quadratic inequalities   34
  2.5 Discriminant of a quadratic equation   35

CHAPTER 3   Indices and surds   43
  3.1 Simplifying expressions involving indices   43
  3.2 Solving equations involving indices   45
  3.3 Simplifying expressions involving surds   47
  3.4 Rationalizing the denominator   48
  3.5 Solving equations involving surds   49

CHAPTER 4   Factors of polynomials   57
  4.1 Operations with polynomials   57
  4.2 Finding zeros of a polynomial function   59
  4.3 The remainder theorem   60
  4.4 The factor theorem   62
  4.5 Rational zeros theorem   64
  4.6 Graphing cubic functions   67
  4.7 Solving cubic inequalities graphically   68

CHAPTER 5   Logarithmic and exponential functions   77
  5.1 Logarithms   77
  5.2 Properties of logarithms   79
  5.3 Solving exponential and logarithmic Equations   82
  5.4 Graphs of logarithmic and exponential functions   84
  5.5 Graphs of $y = ke^{nx} + a$ and $y = k \ln (ax+b)$ 85

CHAPTER 6   Straight-line graphs   95
  6.1 Coordinate geometry   95
  6.2 Finding areas of polygons using shoelace method   98
  6.3 Linear law   100

CHAPTER 7   Coordinate geometry of circles   109
  7.1 The standard equation of a circle   109
  7.2 Intersection of a circle and a straight line   111
  7.3 The equation of a tangent line to a circle   113
  7.4 Intersection of two circles   114

CHAPTER 8 Trigonometry   119

    8.1 Circular measure   119

    8.2 Finding the exact value of the trigonometric functions 123

    8.3 Graphs of trigonometric functions   128

    8.4 Solving trigonometric equations   134

    8.5 Proving trigonometric identities   136

    8.6 Area of non-right angled triangles   138

    8.7 Solving triangles using the law of sines and cosines 139

CHAPTER 9 The binomial theorem   155

    9.1 The Fundamental Counting Principle   155

    9.2 Permutation and combination   157

    9.3 The binomial theorem   159

CHAPTER 10 Sequence and series   169

    10.1 Sequence   169

    10.2 Series   172

CHAPTER 11 Vectors   181

    11.1 Vector notation   181

    11.2 Algebraic operations on vectors   183

    11.3 Vector geometry   184

    11.4 Constant velocity problems   191

CHAPTER 12 Derivative functions   199

    12.1 Instantaneous rate of change   199

    12.2 Finding the derivative functions   202

    12.3 Tangent and normal lines   204

CHAPTER 13 Differentiation rules   211

    13.1 The product and quotient rules   211

    13.2 The chain rule   214

    13.3 The second derivative   217

CHAPTER 14 Applications of differentiation   225

    14.1 Small increments and approximations   225

    14.2 Related rates   228

    14.3 Understanding a curve from the first and second derivatives 231

    14.4 Local maximum and local minimum   233

    14.5 Practical maximum and minimum problems   236

CHAPTER 15 Integration   245

    15.1 Indefinite integrals   245

    15.2 The U-Substitution rule   250

    15.3 Definite integrals   254

    15.4 Area between two curves   257

    15.5 Kinematics   260

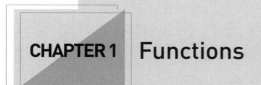 

CHAPTER 1 Functions

1.1 Definition of a function

Definition of a Function

A relation is a set of ordered pairs. A relation or a function can be represented using a mapping diagram which shows how in a relation the elements of the first set (inputs, or preimages) are mapped onto the elements of the second set (outputs or images). The mapping diagrams below show **four different types of relations**: one-to-one, many-to-one, many-to-many, and one-to-many.

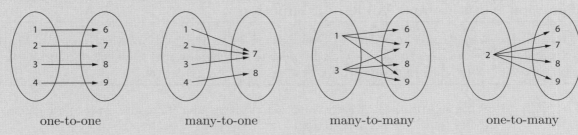

one-to-one   many-to-one   many-to-many   one-to-many

The first mapping diagram represents a one-to-one relation where preimages are $\{1, 2, 3, 4\}$ and images are $\{6, 7, 8, 9\}$. Each image is 5 more than preimage. It can be denoted by $x \mapsto x + 5$.

A function is a special relation where each x-value is related to exactly one corresponding y-value, or where x-values are not repeated. For instance, a set $A = \{(-1, -2), (0, -1), (2, 1), (4, 3)\}$ is a relation because set A is a set of ordered pairs. Also, set A is a function because the x-values, -1, 0, 2, and 4, are not repeated. The set of the x-values and the y-values in set A are called the **domain** and the **range** of the function, respectively. Thus, the domain of the function is $\{-1, 0, 2, 4\}$ and the range of the function is $\{-2, -1, 1, 3\}$.

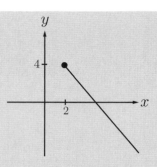

The domain of the graph of a function shown above is $\{x|x \geq 2\}$ or $[2, \infty)$, and Range is $\{y|y \leq 4\}$ or $(-\infty, 4]$.

 Tip
1. One-to-one and many-to-one relations are functions.

2. Many-to-many and one-to-many relations are **not** functions.

3. If set $B = \{(1,3), (2,5), (1,2), (3,5)\}$, set B is a relation, but not a function because the x-value, 1, are repeated twice.

Function Notation and Value of a Function

Functions are often denoted by f or g. The function can be written as $f : x \mapsto f(x)$ or $y = f(x)$. The value of function f at x is denoted by $f(x)$. In order to evaluate the value of function f at $x = 2$, $f(2)$, substitute 2 for x. For instance, if $f(x) = x^2 + 1$, $f(2) = 2^2 + 1 = 5$.

Tip
1. When substituting a negative numerical value in a function, make sure to use parentheses to avoid a mistake. If $f(x) = x^2 + 1$, $f(-2) = (-2)^2 + 1 = 5$.

2. To evaluate $f(x+1)$ when $f(x) = x^2 + 1$, substitute $x + 1$ for x in $f(x)$.

$$f(x+1) = (x+1)^2 + 1 = x^2 + 2x + 2$$

3. $f : x \mapsto x^2 + 1$ is read as 'the function f is such that x is mapped to $x^2 + 1$'.

Vertical Line Test

The vertical line test is a graphical way to determine if a graph represents a function or not. If all vertical lines intersect the graph at most one point, the graph in Figure 1 represents a function. Otherwise, the graph does not represent a function as shown in Figure 2.

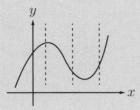

Figure 1

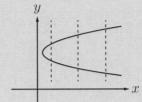

Figure 2

Classification of Functions

- A function $f : A \mapsto B$ is **one-to-one** if every element in the range is the image of exactly one element in the domain. This means that every element in the domain has a different element in the range. If all possible horizontal lines $y = c$ intersect the graph of f at exactly one point, f is one-to-one.

- A function is **many-to-one** if more than one element of the domain has the same image. If any horizontal line $y = c$ intersects the graph of f at two or more points, then f is many-to-one.

 1. The function $f(x) = x^3$ is a one-to-one function.

2. The function $f(x) = x^2 - 1$ is a many-to-one function, neither a one-to-one function.

3. $x = y^2$ or $y = \pm\sqrt{x}$ is a one-to-many mapping, which is **NOT** a function.

Odd and Even Functions

Let $f(x)$ be a function. $f(x)$ is an odd function when $f(-x) = -f(x)$ for all values of x. Any odd function is symmetric with respect to the origin. Whereas, $f(x)$ is an even function when $f(-x) = f(x)$ for all values of x. Any even function is symmetric with respect to the y-axis.

	Odd functions	Even functions
Definition	$f(-x) = -f(x)$	$f(-x) = f(x)$
Graph	Symmetric with respect to the origin	Symmetric with respect to the y-axis
Example	x^3, $\sin x$, $\tan x$	x^2, $\cos x$

Tip   Not all functions are either odd or even. Some functions are neither.

1.2 Composition functions

Composition Functions

If we are given two functions $f(x)$ and $g(x)$, it is possible to create a new function $h(x)$ by composing one into the other.

$$h(x) = fg(x) = (f \circ g)(x) = f(g(x))$$

- The composite function $fg(x)$ or $f(g(x))$ is read as 'f of g of x'.
- $fg(x)$ or f(g(x)) means that the function g acts on x first, then f acts on the result, $g(x)$.
- The domain of $h(x) = f(g(x))$ is the set of all x-values such that x is in the domain of g and $g(x)$ is in the domain of f.
- $f^2(x) = ff(x) = f(f(x))$, which means you apply the function f twice.
- In general, $fg(x) \neq gf(x)$.

Example 1   Finding the composition functions

Let $f(x) = 2x^{-1}$ and $g(x) = x + 1$. Find the following:

(a) $fg(2)$

(b) $fg(x)$

(c) $gf(x)$

(d) $f^2(x)$

(e) The domain of $fg(x)$.

Solution

(a) Since $g(2) = 2 + 1 = 3$, $fg(2) = f(g(2)) = f(3) = \dfrac{2}{3}$.

(b) $fg(x) = f(x + 1) = \dfrac{2}{x + 1}$

(c) $gf(x) = g(f(x)) = g(\frac{2}{x}) = \dfrac{2}{x} + 1$

(d) $f^2(x) = f(f(x)) = f(\frac{2}{x}) = \dfrac{2}{\frac{2}{x}} = x$

(e) The domain of $fg(x)$ consists al all real numbers except $x = -1$ because $g(-1)$ is not in the domain of f.

1.3 Modulus or absolute value functions

Modulus or Absolute Value Functions

The modulus or absolute value function is defined by

$$|x| = \begin{cases} x, & x \geq 0 \\ -x, & x < 0 \end{cases}$$

and the graph is V-shaped.

The graph of the modulus function $y = a|x - h| + k$ has the following characteristics.

- The graph has vertex (h, k) and is symmetric with respect to the line $x = h$.

- If $a > 0$, it opens up.

- If $a < 0$, it opens down.

Properties of Modulus Functions

1. $|x| \geq 0$

2. $|-x| = |x|$

3. $\left||x|\right| = |x|$

4. $|xy| = |x||y|$

5. $\left|\dfrac{x}{y}\right| = \dfrac{|x|}{|y|}$ if $y \neq 0$

6. $|x + y| \leq |x| + |y|$

7. $|x - y| = 0$ if $x = y$

Example 2   Graphing an modulus function

Graph $y = -|2x + 6| + 2$

Solution

$$y = -|2x + 6| + 2$$
$$y = -|2(x + 3)| + 2 \qquad \text{Use } |xy| = |x||y|$$
$$y = -|2||x + 3| + 2$$
$$y = -2|x + 3| + 2$$

The vertex of the modulus function is at $(-3, 2)$ as shown below.

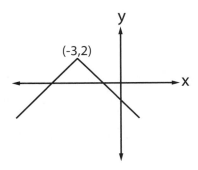

Since $a = -2 < 0$, the graph opens down.

Example 3   Solving a modulus equation

Solve $|x^2 - 6| = 2$

Solution

$$x^2 - 6 = 2 \qquad \text{or} \qquad x^2 - 6 = -2$$
$$x^2 = 8 \qquad \text{or} \qquad x^2 = 4$$
$$x = \pm 2\sqrt{2} \qquad \text{or} \qquad x = \pm 2$$

Since $x = 2\sqrt{2}, -2\sqrt{2}, 2, -2$ satisfy the original equation, the solutions to the equation are either $x = 2\sqrt{2}, -2\sqrt{2}, 2$ or $x = 2$.

 After solving a modulus equation, you always check your answers to make sure that they satisfy the original equation.

1.4 Inverse functions

Inverse Functions

A one-to-one function f maps A onto B will have an inverse function f^{-1} that maps B onto A. If $f(x) = y$, then $f^{-1}(y) = x$.

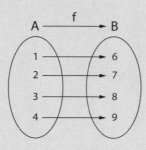

Figure 3

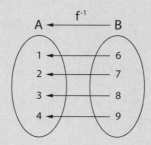

Figure 4

f in Figure 3 is a one-to-one function that maps A onto B. The inverse function in Figure 4, f^{-1}, is a function that maps every element of B onto its corresponding element from A. For instance, $f(2) = 7$. Thus, $f^{-1}(7) = 2$. The meaning of $f^{-1}(7) = 2$ is that the corresponding element of A that f^{-1} maps 7 from B onto is 2.

- The domain of the inverse function f^{-1} is the range of the function f.

- The range of the inverse function f^{-1} is the domain of the function f.

Horizontal Line Test

The horizontal line test is a graphical way to determine if a function has an inverse function or not. If all horizontal lines intersect the graph at exactly one point, the function has an inverse function as shown in Figure 5. Otherwise, the function does not have an inverse function as shown in Figure 6.

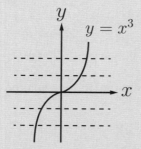

Figure 5: f has an inverse function

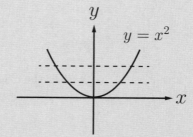

Figure 6: f does not have an inverse function

Example 4   Informal concept of an inverse function

If $f(x) = \sqrt{x-1}$, find $f^{-1}(2)$.

Solution $f(5) = \sqrt{5-1} = \sqrt{4} = 2$. Since $f^{-1}(2)$ means to find the value of x such that $f(x) = 2$. Since $f(5) = 2$, $x = f^{-1}(2) = 5$.

Finding Inverse Functions

There are four steps to find the inverse of a function.

    Step 1:   Replace $f(x)$ with the y variable.

    Step 2:   Switch the x and y variables.

    Step 3:   Solve for y.

    Step 4:   [Optional] Replace the y variable with $f^{-1}(x)$.

 1. Perform the horizontal line test to determine if a function has an inverse function. If the function passes the horizontal line test, follow the four steps shown above to find the inverse function.

    2. The inverse of f is denoted by f^{-1}. Note that the -1 used in f^{-1} is not an exponent. Thus, the inverse function of $f(x)$, $f^{-1}(x)$, is not the reciprocal of $f(x)$. In other words, $f^{-1}(x) \neq \dfrac{1}{f(x)}$. For instance, if $f(x) = x^3 + 5$, $f^{-1}(x) \neq \dfrac{1}{x^3 + 5}$.

    3. $ff^{-1}(x) = f^{-1}f(x) = x$

    4. $(f \circ g)^{-1} = g^{-1} \circ f^{-1}$

Example 5   Finding the inverse function

Find the inverse function of $f(x) = x^3 + 5$.

Solution   Replace $f(x)$ with the y variable so that $f(x) = x^3 + 5$ becomes $y = x^3 + 5$. Next, switch the x and y variables and solve for y.

$$y = x^3 + 5 \qquad \text{Switch the } x \text{ and } y \text{ variables}$$
$$x = y^3 + 5 \qquad \text{Subtract 5 from each side}$$
$$x - 5 = y^3 \qquad \text{Take the cube root of each side}$$
$$y = \sqrt[3]{x-5}$$

Therefore, the inverse function of $f(x) = x^3 + 5$ is $y = \sqrt[3]{x-5}$.

Graphing the Inverse Functions

The graph of the function $f(x) = \sqrt{x} + 2$ is shown in Figure 7. The graph of the inverse function is obtained by reflecting the graph of $f(x)$ about the line $y = x$ as shown in Figure 8. In other words, the graph of $f(x)$ and its inverse function $f^{-1}(x)$ are symmetric with respect to the line $y = x$.

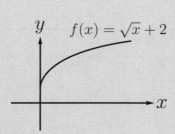

Figure 7: The graph of $f(x)$

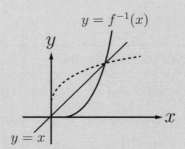

Figure 8: The graph of $f^{-1}(x)$

Example 6   Finding the inverse function

Find the inverse function of $f(x) = \sqrt{x} + 2$ and state the domain and range of the inverse function.

Solution   The graph of $f(x) = \sqrt{x} + 2$ shown in Figure 7 suggests that the domain and range of the function are $x \geq 0$ and $y \geq 2$, respectively. Since the domain of a function is the range of the inverse function, and the range of a function is the domain of the inverse function, the domain and range of the inverse function are $x \geq 2$ and $y \geq 0$, respectively. The graph of the inverse function is shown in Figure 8.

In order to find the inverse function algebraically, switch the x and y variables and solve for y.

$$y = \sqrt{x} + 2 \qquad \text{Switch the } x \text{ and } y \text{ variables}$$
$$x = \sqrt{y} + 2 \qquad \text{Subtract 2 from each side}$$
$$x - 2 = \sqrt{y} \qquad \text{Square both sides}$$
$$y = (x - 2)^2 \qquad \text{Replace } y \text{ with } f^{-1}(x)$$
$$f^{-1}(x) = (x - 2)^2, \quad x \geq 2 \qquad \text{Since the domain of } f^{-1}(x) \text{ is } x \geq 2$$

Therefore, the inverse function is $f^{-1}(x) = (x - 2)^2$, $x \geq 2$.

1.5 Transformations

Graphs of Parent Functions

A family of functions is a group of functions that all have a similar shape. A parent function is the simplest function and is used as a reference to graph more complicated functions in the family. The table below summarizes the graph, domain, and range of each parent function.

Parent function	Graph	Parent function	Graph
Constant $y = k$ Domain: $(-\infty, \infty)$ Range: $y = k$		Greatest integer $y = [x]$ Domain: $(-\infty, \infty)$ Range: All integers	
Linear $y = x$ Domain: $(-\infty, \infty)$ Range: $(-\infty, \infty)$		Absolute value $y = |x|$ Domain: $(-\infty, \infty)$ Range: $[0, \infty)$	
Quadratic $y = x^2$ Domain: $(-\infty, \infty)$ Range: $[0, \infty)$		Square root $y = \sqrt{x}$ Domain: $[0, \infty)$ Range: $[0, \infty)$	
Cubic $y = x^3$ Domain: $(-\infty, \infty)$ Range: $(-\infty, \infty)$		Cube root $y = \sqrt[3]{x}$ Domain: $(-\infty, \infty)$ Range: $(-\infty, \infty)$	
Rational 1 $y = \frac{1}{x}$ Domain: $x \neq 0$ Range: $y \neq 0$		Rational 2 $y = \frac{1}{x^2}$ Domain: $x \neq 0$ Range: $(0, \infty)$	
Exponential $y = a^x, a > 1$ Domain: $(-\infty, \infty)$ Range: $(0, \infty)$		Logarithmic $y = \log_a x, a > 1$ Domain: $(0, \infty)$ Range: $(-\infty, \infty)$	

Transformations

The general shape of each parent function can be moved or resized by transformations. For instance, the three functions shown below

$$y = x^2, \qquad y = (x-1)^2, \qquad y = \frac{1}{2}x^2 + 3$$

are in the family of quadratic functions and have the same shape. After moving or resizing the graph of the parent function, $y = x^2$, we can obtain the graphs of $y = (x-1)^2$ and $y = \frac{1}{2}x^2 + 3$.

Transformations consist of horizontal and vertical shifts, horizontal and vertical stretches and compressions, and reflections about the x-axis and y-axis. The table below summarizes the transformations.

Transformation	Function Notation	Effect on the graph of $f(x)$
Horizontal shift	$y = f(x-1)$	Move the graph of $f(x)$ right 1 unit
	$y = f(x+2)$	Move the graph of $f(x)$ left 2 units
Vertical shift	$y = f(x) + 3$	Move the graph of $f(x)$ up 3 units
	$y = f(x) - 4$	Move the graph of $f(x)$ down 4 units
Horizontal stretch and compression	$y = f(2x)$	Horizontal compression of the graph of $f(x)$ by a factor of $\frac{1}{2}$
	$y = f(\frac{1}{3}x)$	Horizontal stretch of the graph of $f(x)$ by a factor of 3
Vertical stretch and compression	$y = 3f(x)$	Vertical stretch of the graph of $f(x)$ by a factor of 3
	$y = \frac{1}{2}f(x)$	Vertical compression of the graph of $f(x)$ by a factor of $\frac{1}{2}$
Reflection about the x-axis, y-axis, and origin	$y = -f(x)$	Reflect the graph of $f(x)$ about the x-axis
	$y = f(-x)$	Reflect the graph of $f(x)$ about the y-axis
	$y = -f(-x)$	Reflect the graph of $f(x)$ about the origin

 1. Translation means moving right, left, up, or down. Sometimes, horizontal shifts or vertical shifts are referred to as translations.

2. Horizontal shifts, written in the form $f(x-h)$, do the opposite of what they look like they should do. $f(x-1)$ means to move the graph of $f(x)$ right 1 unit. Whereas, $f(x+2)$ means to move the graph of $f(x)$ left 2 units.

3. Horizontal stretches and compressions, written in the form $f(cx)$, also do the opposite of what they look like they should do. $f(2x)$ means a horizontal compression of the graph of $f(x)$ by a factor of $\frac{1}{2}$. Whereas, $f(\frac{1}{3}x)$ means a horizontal stretch of the graph of $f(x)$ by a factor of 3.

Order of Transformations

In order to graph a function involving more than one transformation, use the order of transformations.

1. Horizontal shift

2. Horizontal and vertical stretch and compression

3. Reflection about x-axis and y-axis

4. Vertical shift

The order of transformations suggests to first perform the horizontal shift. Afterwards, perform the horizontal and vertical stretch and compression. Next, perform the reflection about the x-axis and y-axis. Finally, perform the vertical shift.

Example 7   Graphing the function by transformations

Graph $f(x) = -\sqrt{x-1} + 2$ and state the domain and range of the function.

Solution   If $f(x) = \sqrt{x}$, $-f(x-1) + 2$ in function notation represents $-\sqrt{x-1} + 2$. The function notation $-f(x-1) + 2$ suggests that the graph of $-\sqrt{x-1} + 2$ involves a horizontal shift, reflection about the x-axis, and a vertical shift from the graph of the parent function $y = \sqrt{x}$. Thus, perform the horizontal shift first. Next, perform the reflection about the x-axis, and lastly, perform the vertical shift.

Let's start with the graph of $f(x) = \sqrt{x}$.

$$f(x) = \sqrt{x}$$ The graph of the parent function as shown in Fig. 1

$$f(x-1) = \sqrt{x-1}$$ Move the graph of $\sqrt{x}$ right 1 unit as shown in Fig. 2

$$-f(x-1) = -\sqrt{x-1}$$ Reflect the graph of $\sqrt{x-1}$ about the x-axis as shown in Fig. 3

$$-f(x-1) + 2 = -\sqrt{x-1} + 2$$ Move the graph of $-\sqrt{x-1}$ up 2 units as shown in Fig. 4

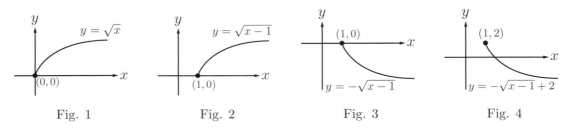

Fig. 1   Fig. 2   Fig. 3   Fig. 4

As shown in Fig. 4, the domain of the function is $[1, \infty)$, and the range of the function is $(-\infty, 2]$.

Graph of $y = |f(x)|$

In order to graph the function $y = |f(x)|$, reflect in the x-axis the part of the graph of $y = f(x)$ that lies below the x-axis. For instance, in order to graph $y = |x^2 - 3|$, start with the graph of $y = x^2 - 3$ as shown in Figure 5. Determine the part of the graph that lies below the x-axis as shown in Figure 6. Lastly, reflect in the x-axis the part of the graph that lies below the x-axis as shown in Figure 7.

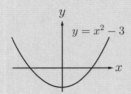

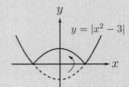

Fig 5: $y = x^2 - 3$ Fig 6: Part below the x-axis   Fig 7: $y = |x^2 - 3|$

Tip   In general, the graph of $y = |f(x)|$ is **NOT** a reflection of the entire graph of $y = f(x)$ about the x-axis nor about the y-axis.

Example 8   Graphing $y = |f(x)|$

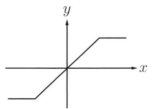

The graph of $y = f(x)$ is given above. Graph $y = |f(x)|$.

Solution   The function $f(x)$ is an odd function since the graph of $y = f(x)$ is symmetric with respect to the origin. To graph $y = |f(x)|$, reflect in the x-axis the part of the graph of $y = f(x)$ that lies below the x-axis. The graph of $y = |f(x)|$ is shown below.

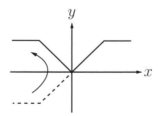

Graph of $y = |f(x)|$

Since the graph of $y = |f(x)|$ is symmetric with respect to the y-axis, the function $y = |f(x)|$ is an even function.

EXERCISES

1. The function f is defined by $f : x \mapsto \dfrac{3x+1}{2}$. Find $f^{-1}(5)$.

2. The function f is defined by $f : x \mapsto |3x - 6|$ for $x \in \mathbb{R}$.

   (a) Sketch the graph of $y = f(x)$.

   (b) Solve $|3x - 6| = 7$.

   (c) Solve $|3x - 6| > 7$.

3. A function f is such that $f(x) = x^2 - 6x + 5$ for $-1 \leq x \leq 6$.

   (a) Find the range of f.

   (b) Write down a suitable domain of f for which f^{-1} exists.

4. Suppose $f(x) = |2x + 1|$ and $g(x) = |x - 2|$.

   (a) Sketch the graph of $y = f(x)$ and $y = g(x)$.

   (b) Solve $|2x + 1| = |x - 2|$.

   (c) Solve $|2x + 1| < |x - 2|$.

5. The function f is defined by $f(x) = \sqrt{x+2} + 3$ for $x \geq -2$.

    (a) Find the expression for $f^{-1}(x)$.

    (b) Solve the equation $f^{-1} = f(23)$.

6. The function f is defined by $f(x) = |x^2 - 4x + 3|$ for $x \in \mathbb{R}$.

    (a) Sketch the graph of $y = f(x)$.

    (b) Find the values of x for which $|x^2 - 4x + 3| = 3 - x$.

7. The function f is defined by $f : x \mapsto 3x^2 + 1$ for real values of x.

   (a) Express $f(x+1)$ in the form $ax^2 + bx + c$.

   (b) Solve $f(x+1) + f(x-1) = 20$.

8. The functions f and g are defined by $f(x) = 2x - 1$ and $g(x) = x^2 - 6$.

   (a) Evaluate $gf(-1)$

   (b) Evaluate $g^2(-1)$

   (c) Find the values of x for which $gf(x) = x + 3$.

9. Suppose $f(x) = 4x - 6$ and $g(x) = \dfrac{kx}{2} + 1$.

   (a) Find $ff(x)$ in terms of x.

   (b) Given that $fg(6) = 12$, find the value of k.

10. The function f is defined by $f(x) = \dfrac{2}{x+2} - \dfrac{1}{x-1}$.

   (a) Find the values of x for which $f(x) = 0$.

   (b) Find the values of x for which $f(x)$ is not defined.

   (c) Find the possible values of x for which $f(x) = 3$. Give your answer in the form $\dfrac{p \pm \sqrt{q}}{r}$ where, p, q, and r are positive integers.

11. The functions f and g are defined, for $x \in \mathbb{R}$, by

$$f : x \mapsto x + 2$$
$$g : x \mapsto \frac{x - 1}{x + 2}$$

(a) Find $g^2(x)$ in terms of x.

(b) Find gf in terms of x.

(c) Find fg in terms of x

(d) Find $(fg)^{-1}$.

CHAPTER 2 | Quadratics

2.1 Quadratic functions

Quadratic Functions

A quadratic function is a polynomial function of degree 2 and can be expressed in three forms.

Standard form:	$y = ax^2 + bx + c$	
Vertex form:	$y = a(x - h)^2 + k,$	Vertex (h, k)
Factored form:	$y = a(x - p)(x - q),$	where p and q are x-intercepts.

The graph of a quadratic function is a parabola. Depending on the value of the **leading coefficient**, $a\,(a \neq 0)$, the graph of a quadratic function either opens up or opens down.

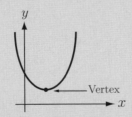

Figure 1: If $a > 0$, opens up

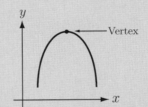

Figure 2: If $a < 0$, opens down

The vertex is the minimum point when the parabola opens up as shown in Figure 1 and the maximum point when the parabola opens down as shown in Figure 2.

The x-coordinate of the vertex from each form are as follows:

$$\text{Standard form: } y = ax^2 + bx + c \implies x = -\frac{b}{2a}$$

$$\text{Vertex form: } y = a(x - h)^2 + k \implies x = h$$

$$\text{Factored form: } y = a(x - p)(x - q) \implies x = \frac{p + q}{2}$$

To evaluate the y-coordinate of the vertex, substitute the value of the x-coordinate in each form.

The x-intercept and y-intercept of quadratic functions

- To solve for the x-intercept: substitute 0 for y. Then, quadratic functions become quadratic equations as shown below.

$$\text{Standard form: } y = ax^2 + bx + c \implies ax^2 + bx + c = 0$$

$$\text{Vertex form: } y = a(x - h)^2 + k \implies a(x - h)^2 + k = 0$$

$$\text{Factored form: } y = a(x - p)(x - q) \implies a(x - p)(x - q) = 0$$

- To solve for the y-intercept: substitute 0 for x and evaluate the y-intercept.

2.2 Maximum and minimum value by completing the square

Completing the Square

Completing the square is a technique for converting a quadratic of the form $ax^2 + bx + c$ to the form $a(x-h)^2 + k$ for some values of h and k. In other words, completing the square places a perfect square trinomial inside of a quadratic expression. For instance, to complete the square for $x^2 + bx$, you need to add $\left(\dfrac{b}{2}\right)^2$ to make $\left(x + \dfrac{b}{2}\right)^2$.

$$x^2 + bx + \left(\frac{b}{2}\right)^2 \quad \longrightarrow \quad \left(x + \frac{b}{2}\right)^2$$

Completing the square is used in finding the maximum and minimum points of quadratic functions which are also called **turning points** or **stationary points**.

Example 1 Finding the maximum or minimum by completing the square

(a) Express $2x^2 - 12x + 22$ in the form $a(x-h)^2 + k$, where a, h and k are constants.

(b) Find the maximum or minimum value from the vertex form $f(x) = a(x-h)^2 + k$.

Solution

(a)

$$\begin{aligned}
2x^2 - 12x + 22 &= 2(x^2 - 6x + 0) + 22 \\
&= 2(x^2 - 6x + 9 - 9) + 22 \\
&= 2(x^2 - 6x + 9) - 18 + 22 \\
&= 2(x-3)^2 + 4
\end{aligned}$$

(b) The vertex is located at $(3, 4)$. Since the leading coefficient of $2(x-3)^2 + 4$ is positive, the graph of $f(x)$ opens up. Thus, the minimum value of $f(x)$ is 4.

2.3 Solving quadratic equations

Solving Quadratic Equations

Solving a quadratic equation is finding the x-intercept(s) of the quadratic function. There are three common methods to solve a quadratic equation: **Factoring**, **Completing the square**, and **Quadratic formula**.

1. Factoring

Factoring is an important tool that is required for solving a quadratic equation. Factoring is the opposite of expanding. Factoring a quadratic expression is to write the expression as a product of two linear terms. Below is an example.

$$(x-2)(x-3) \quad \xrightarrow{\text{Expanding}} \quad x^2 - 5x + 6$$
$$x^2 - 5x + 6 \quad \xrightarrow{\text{Factoring}} \quad (x-2)(x-3)$$

If a quadratic equation can be expressed as $x^2 + (p+q)x + pq = 0$, it can be factored as $(x+p)(x+q) = 0$. For instance,

$$x^2 - 5x + 6 = 0 \quad \Longrightarrow \quad x^2 + (-2 + -3)x + (-2)(-3) = 0 \quad \Longrightarrow \quad (x-2)(x-3) = 0$$

Once a quadratic equation is written in a factored form, use the **zero product property** to solve the equation.

$$\text{Zero product property:} \qquad \text{If } ab = 0, \text{ then } a = 0 \text{ or } b = 0$$

Thus, the solutions to $(x-2)(x-3) = 0$ is

$$(x-2)(x-3) = 0 \quad \Longrightarrow \quad (x-2) = 0 \quad \text{or} \quad (x-3) = 0 \quad \Longrightarrow \quad x = 2 \quad \text{or} \quad x = 3$$

2. Completing the square

Equations of the form $ax^2 + bx + c = 0$ can be converted to the form $(x+p)^2 = q$ from which the solutions are easy to obtain. For instance,

$$x^2 + 2x - 2 = 0 \quad \Longrightarrow \quad x^2 + 2x = 2 \quad \Longrightarrow \quad x^2 + 2x + 1 = 3 \quad \Longrightarrow \quad (x+1)^2 = 3$$

3. The Quadratic Formula

The quadratic formula is a general formula for solving quadratic equations. The solutions to the quadratic equation $ax^2 + bx + c = 0$ are as follows:

$$x = \frac{-b \pm \sqrt{b^2 - 4ac}}{2a}$$

Example 2   Finding solutions of a quadratic equation

Solve the equation: $x^2 - x - 1 = 0$

Solution   Since the quadratic equation cannot be factored, use the quadratic formula to solve the equation.

$$
\begin{aligned}
x &= \frac{-b \pm \sqrt{b^2 - 4ac}}{2a} \\
&= \frac{-(-1) \pm \sqrt{(-1)^2 - 4(1)(-1)}}{2(1)} \\
&= \frac{1 \pm \sqrt{5}}{2}
\end{aligned}
$$

Substitute 1 for a, -1 for b, and -1 for c

2.4 Solving quadratic inequalities

Solving Quadratic Inequalities

Solving a quadratic inequality means finding the x-values for which the graph of a quadratic function lies above or below the x-axis. A quadratic inequality can be solved algebraically. However, solving a quadratic inequality **graphically** is highly recommended.

In order to solve $ax^2 + bx + c > 0$ (or $ax^2 + bx + c \geq 0$) graphically,

Step 1 Find the x-intercepts of $y = ax^2 + bx + c$: Let $y = 0$ and solve for x using factoring or the quadratic formula.

Step 2 Graph $y = ax^2 + bx + c$.

Step 3 From the graph in step 2, find the x-values for which the graph lies **above** (or on and above) the x-axis.

In order to solve $ax^2 + bx + c < 0$ (or $ax^2 + bx + c \leq 0$) graphically,

Step 1 Find the x-intercepts of $y = ax^2 + bx + c$: Let $y = 0$ and solve for x using factoring or the quadratic formula.

Step 2 Graph $y = ax^2 + bx + c$.

Step 3 From the graph in step 2, find the x-values for which the graph lies **below** (or on and below) the x-axis.

Example 3   Solving a quadratic inequality graphically

Solve $x^2 - 5x + 6 \geq 0$

Solution   Substitute 0 for y in $y = x^2 - 5x + 6$ and solve $x^2 - 5x + 6 = 0$ using factoring.

$$(x - 2)(x - 3) = 0 \implies (x - 2) = 0 \quad \text{or} \quad (x - 3) = 0 \implies x = 2 \quad \text{or} \quad x = 3$$

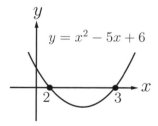

Figure 3

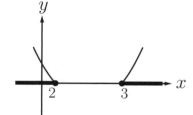

Figure 4

As shown in Figure 3, the x-intercepts of $y = x^2 - 5x + 6$ are 2, and 3. Since the graph lies on and above the x-axis to the left of $x = 2$ and to the right of $x = 3$ as shown in Figure 4, the solution to $x^2 - 5x + 6 \geq 0$ is $x \leq 2$ or $x \geq 3$.

2.5 Discriminant of a quadratic equation

The Discriminant of a Quadratic Equation $ax^2 + bx + c = 0$

In the quadratic formula $x = \dfrac{-b \pm \sqrt{b^2 - 4ac}}{2a}$, the quantity $b^2 - 4ac$ under the square root sign is called **discriminant**. The symbol **delta** Δ is used to represent the discriminant.

- If $\Delta > 0$, there are two distinct real roots.

- If $\Delta = 0$, there are two real equal roots.

- If $\Delta < 0$, there are no real roots.

Intersection of a Line and a Curve

When a line $y = mx + b$ intersects a curve $y = ax^2 + px + q$, set up the equation $ax^2 + px + q = mx + b$ and get the form $ax^2 + bx + c = 0$. The solutions to $ax^2 + bx + c = 0$ are the x-coordinates of the intersection points of the line and the curve.

Use the discriminant to find the number of intersections points of the line and the curve.

- If $\Delta > 0$, there are two distinct intersection points.

- If $\Delta = 0$, there is only one intersection point (the line is tangent to the curve)

- If $\Delta < 0$, there are no intersection points.

Example 4 Finding the range of values of k using discriminant

Find the range of values of k for which $2x^2 + (k-2)x + 18 = 0$ has no real roots.

Solution For no real roots, $b^2 - 4ac < 0$.

$$b^2 - 4ac < 0$$
$$(k-2)^2 - 4(2)(18) < 0$$
$$(k-2)^2 < 12^2$$
$$-12 < k - 2 < 12$$
$$-10 < k < 14$$

Example 5   Finding the value of k using discriminant

Find the value of k for which $y = -x + k$ is a tangent to the curve $y = x^2 - 4x + 3$.

Solution   Set up the equation $x^2 - 4x + 3 = -x + k$ and get the form $x^2 - 3x + (3 - k) = 0$. Since the line is a tangent to the curve,

$$b^2 - 4ac = 0$$
$$(-3)^2 - 4(1)(3 - k) = 0$$
$$9 - 12 + 4k = 0$$
$$4k = 3$$
$$k = \frac{3}{4}$$

Example 6   Finding the range of values of k using discriminant

Find the range of values of k for which the $y = kx + 2$ does not intersect the curve $y = 4x^2 - 2x + 3$.

Solution   Set up the equation $4x^2 - 2x + 3 = kx + 2$ and get the form $4x^2 - (k + 2)x + 1 = 0$. Since the line and the curve do not intersect,

$$b^2 - 4ac < 0$$
$$(k + 2)^2 - 4(4)(1) < 0$$
$$k^2 + 4k - 12 < 0$$
$$(k + 6)(k - 2) < 0$$
$$-6 < k < 2$$

EXERCISES

1. Joshua is using the quadratic formula to solve a quadratic equation. He substitutes values into the formula and gets

$$x = \frac{5 \pm \sqrt{25 - 12}}{6}$$

Find the quadratic equation that Joshua is solving in the form $ax^2 + bx + c$, where a, b, and c are integers.

2. The expression $x^2 + 4mx$ has a minimum value as x varies.

(a) Find the minimum value of $x^2 + 4mx$.

(b) Find the value of x for which this minimum value occurs.

3. A store sold T-shirts at low wholesale price. If the price of a T-shirt was \$12, the store sold 240 T-shirts. The store realized that for every 50 cents reduction on the price of a T-shirt, the store sold 40 more T-shirts.

   (a) Write a quadratic model that represents the revenue of the store.

   (b) What price of a T-shirt will maximize the revenue of the store?

   (c) How many T-shirts will be sold at the price obtained in part (b)?

   (d) What is the maximum revenue of the store?

4. The area of a square with side length x is equal to the perimeter of an equilateral triangle with side length $x + 6$. Find the area of the equilateral triangle.

5. A projectile is launched at 19.6 meters per second (m/s) from a 102.9-meter tall building. The quadratic model for the projectile's height h, in meters, at time x seconds after launch is given by

$$h(x) = -4.9x^2 + 19.6x + 102.9, \text{ where } x \geq 0$$

   (a) Find the projectile's height at 3 seconds after launch.

   (b) When does the projectile reach its maximum height?

   (c) When does the projectile hit the ground?

6. There are k jelly beans in a jar. 4 of the jelly beans are red and the rest of the jelly beans are blue. Jason takes a jelly bean at random from the jar and eats it. Jason then takes another jelly bean at random from the jar and eats it. The probability that Jason eats two red jelly beans is $\frac{2}{5}$. find the number of jelly beans k in the jar.

   (a) Show that the probability that Jason eats two red jelly beans is $\frac{2}{5}$ results in a quadratic equation $k^2 - k = 30$.

   (b) Find the total number of jelly beans k in the jar.

7. Solve the following equations. Give your solutions correct to 3 significant figures.

   (a) $x - \dfrac{2}{x} = 3$

   (b) $\dfrac{5}{x^2} + \dfrac{2}{x} = 4$

8. Let $f(x) = \left|(x+2)(x-4)\right|$.

    (a) Sketch the graph of $f(x)$.

    (b) Find the coordinates of the stationary point on the graph of $f(x)$.

    (c) Find the set of values of k for which $|(x+2)(x-4)| = k$ has four solutions.

    (d) Solve $\left|(x+2)(x-4)\right| < 16$.

9. Solve the following quadratic inequalities.

    (a) $x^2 < x + 20$

    (b) $2x^2 \le 5x - 2$

    (c) $3x^2 + 2x > 2x^2 - x + 10$

10. $f(x) = 1 + 2x - 2x^2$ for $x \leq \frac{1}{2}$.

    (a) Express $1 + 2x - 2x^2$ in the form $a - b(x + c)^2$. Find the value of a, b, and c.

    (b) Find the coordinates of the turning point of $f(x)$, stating wether it is a maximum or minimum point.

    (c) Explain why $f(x)$ has an inverse function and find the expression for $f^{-1}(x)$ in terms of x.

11. The line $y = 2x + k$ is a tangent to the curve $x^2 + xy + 12 = 0$.

    (a) Find the possible values of k.

    (b) For $k > 0$, find the coordinates of the point of tangency.

CHAPTER 3 Indices and surds

3.1 Simplifying expressions involving indices

Indices Rules or Properties of Exponents

The table below summarizes the indices rules.

Indices Rules	Example
1. $a^m \cdot a^n = a^{m+n}$	1. $2^4 \cdot 2^6 = 2^{10}$
2. $\frac{a^m}{a^n} = a^{m-n}$	2. $\frac{2^{10}}{2^3} = 2^{10-3} = 2^7$
3. $(a^m)^n = a^{mn} = (a^n)^m$	3. $(2^3)^4 = 2^{12} = (2^4)^3$
4. $a^0 = 1$	4. $(-2)^0 = 1, \ (3)^0 = 1, \ (100)^0 = 1$
5. $a^{-1} = \frac{1}{a}$	5. $2^{-1} = \frac{1}{2}$
6. $a^{\frac{1}{n}} = \sqrt[n]{a}$	6. $2^{\frac{1}{2}} = \sqrt{2}, \quad x^{\frac{1}{3}} = \sqrt[3]{x}$
7. $a^{\frac{m}{n}} = (a^m)^{\frac{1}{n}} = \sqrt[n]{a^m}$	7. $2^{\frac{3}{2}} = \sqrt[2]{2^3}, \quad x^{\frac{3}{4}} = \sqrt[4]{x^3}$
8. $a^{-\frac{m}{n}} = (a^{\frac{m}{n}})^{-1} = \frac{1}{\sqrt[n]{a^m}}$	8. $2^{-\frac{3}{4}} = (2^{\frac{3}{4}})^{-1} = \frac{1}{\sqrt[4]{2^3}}$
9. $(ab)^n = a^n b^n$	9. $(2 \cdot 3)^6 = 2^6 \cdot 3^6, \quad (2x)^2 = 2^2 x^2$
10. $\left(\frac{a}{b}\right)^n = \frac{a^n}{b^n}$	10. $\left(\frac{2}{x}\right)^3 = \frac{2^3}{x^3}$
11. $\frac{b^{-n}}{a^{-m}} = \frac{a^m}{b^n}$	11. $\frac{y^{-3}}{x^{-2}} = \frac{x^2}{y^3}$

Example 1   Simplifying an expression involving indices

Simplify $\left(\dfrac{2x}{3y^2}\right)^{-3}$.

Solution   Let's rewrite $\left(\dfrac{2x}{3y^2}\right)^{-3}$ as $\left(\left(\dfrac{2x}{3y^2}\right)^{-1}\right)^3$ first.

$$\left(\frac{2x}{3y^2}\right)^{-3} = \left(\left(\frac{2x}{3y^2}\right)^{-1}\right)^3$$

$$= \left(\frac{3y^2}{2x}\right)^3$$

$$= \frac{3^3(y^2)^3}{2^3 x^3}$$

$$= \frac{27y^6}{8x^3}$$

Example 2   Simplifying an expression involving indices

Find the values of a and b for the expression below.

$$\frac{\sqrt{x^{-1}} \times \sqrt[3]{y}}{\sqrt{x^4 y^{-\frac{4}{3}}}} = x^a y^b$$

Solution

$$\frac{\sqrt{x^{-1}} \times \sqrt[3]{y}}{\sqrt{x^4 y^{-\frac{4}{3}}}} = \frac{(x^{-1})^{\frac{1}{2}} \times y^{\frac{1}{3}}}{(x^4)^{\frac{1}{2}}(y^{-\frac{4}{3}})^{\frac{1}{2}}}$$

$$= \frac{x^{-\frac{1}{2}} y^{\frac{1}{3}}}{x^2 y^{-\frac{2}{3}}}$$

$$= \left(\frac{x^{-\frac{1}{2}}}{x^2}\right)\left(\frac{y^{\frac{1}{3}}}{y^{-\frac{2}{3}}}\right)$$

$$= x^{-\frac{5}{2}} y$$

Therefore, $a = -\dfrac{5}{2}$ and $b = 1$.

3.2 Solving equations involving indices

Solving Equations involving Indices

Equations that involve terms of the form a^x, $a > 0$, $a \neq 1$, are referred to as **exponential equations**. Such equations can be solved by applying the indices rules.

- When each side of an equation involving indices has the same base:

$$\text{If } a^x = a^y \implies x = y \qquad\qquad \text{e.g. If } 2^x = 2^3, \text{ then } x = 3$$

Example 3   Solving an equation involving Indices

Solve the following equation:

$$2^x \cdot 8^{-x} = 4^{x-3}$$

Solution   Since $8^{-x} = (2^3)^{-x} = 2^{-3x}$, and $4^{x-3} = (2^2)^{x-3} = 2^{2x-6}$

$$2^x \cdot 8^{-x} = 4^{x-3}$$
$$2^x \cdot 2^{-3x} = 2^{2x-6}$$
$$2^{-2x} = 2^{2x-6}$$
$$-2x = 2x - 6$$
$$-4x = -6$$
$$x = \frac{3}{2}$$

Therefore, the solution to the equation $2^x \cdot 8^{-x} = 4^{x-3}$ is $x = \frac{3}{2}$.

Example 4   Solving an equation involving Indices

Solve the following equation:

$$3 \cdot 9^x + 5 \cdot 3^x - 2 = 0$$

Solution   Let $y = 3^x$. 9^x can be written as y^2 as shown below.

$$
\begin{aligned}
9^x &= (3^2)^x \\
&= 3^{2x} \\
&= (3^x)^2 \\
&= y^2
\end{aligned}
$$

Rewrite the equation $3 \cdot 9^x + 5 \cdot 3^x - 2 = 0$ in terms of y as $3y^2 + 5y - 2 = 0$ and solve the equation.

$3 \cdot 9^x + 5 \cdot 3^x - 2 = 0$	Let $y = 3^x$
$3y^2 + 5y - 2 = 0$	Factorize
$(3y - 1)(y + 2) = 0$	
$y = \dfrac{1}{3}$ or $y = -2$	Replace y with 3^x
$3^x = \dfrac{1}{3}$ or $3^x = -2$	

Solving $3^x = \dfrac{1}{3}$ gives $x = -1$. There is no solution to $3^x = -2$, since $3^x > 0$ for all real values of x. Therefore, the solution to the equation $3 \cdot 9^x + 5 \cdot 3^x - 2 = 0$ is $x = -1$.

3.3 Simplifying expressions involving surds

Surds

A **surd** or **radical** is an irrational number of the form $\sqrt{n}$, where n is a positive integer that is not a perfect square. For instance, $\sqrt{2}$, $\sqrt{3}$, $3 - \sqrt{2}$, and $\dfrac{2 + \sqrt{7}}{5}$ are all surds. However, $\sqrt{16}$ is not a surd since $\sqrt{16} = 4$.

Properties of Surds

- Product property: $\sqrt{a} \times \sqrt{b} = \sqrt{ab}$, where $a, b \geq 0$

- Quotient property: $\dfrac{\sqrt{a}}{\sqrt{b}} = \sqrt{\dfrac{a}{b}}$, where $a \geq 0$, and $b > 0$

[Tip] $\quad \sqrt{a} \times \sqrt{a} = \sqrt{a^2} = a$

Example 5   Simplifying expressions involving Surds

Simplify the following expressions.

(a) $3\sqrt{2} \times 4\sqrt{3}$

(b) $\sqrt{8} + \sqrt{18} + \sqrt{32}$

(c) $(3 - \sqrt{2})^2$

(d) $(1 + \sqrt{2})(3 - \sqrt{3})$

Solution

(a) $3\sqrt{2} \times 4\sqrt{3} = 3 \times 4 \times \sqrt{2} \times \sqrt{3} = 12\sqrt{6}$

(b) $\sqrt{8} + \sqrt{18} + \sqrt{32} = 2\sqrt{2} + 3\sqrt{2} + 4\sqrt{2} = 9\sqrt{2}$

(c) $(3 - \sqrt{2})^2 = 3^2 - 2(3)\sqrt{2} + (\sqrt{2})^2 = 11 - 6\sqrt{2}$

(d) $(1 + \sqrt{2})(3 - \sqrt{3}) = 3 - \sqrt{3} + 3\sqrt{2} - \sqrt{6}$

3.4 Rationalizing the denominator

Rationalizing the Denominator of a Fraction

To rationalize the denominator of a fraction means to turn an irrational denominator into a rational number. The **conjugate surds**, $(a + b\sqrt{c}$, and $a - b\sqrt{c})$, are very useful when rationalizing the denominator because the product of two conjugate surds always gives a rational number.

$$\text{Product of two conjugate surds :} \qquad (a + b\sqrt{c})(a - b\sqrt{c}) = a^2 - b^2 c$$

Rationalize the denominator of a fraction using the following rules:

- For a fraction of the form $\dfrac{1}{\sqrt{a}}$: multiply the numerator and denominator by $\sqrt{a}$.

- For a fraction of the form $\dfrac{1}{a + b\sqrt{c}}$: multiply the numerator and denominator by $a - b\sqrt{c}$.

- For a fraction of the form $\dfrac{1}{a - b\sqrt{c}}$: multiply the numerator and denominator by $a + b\sqrt{c}$.

Example 6   Rationalizing the denominator

Rationalize the following expressions.

(a) $\dfrac{3}{-\sqrt{5}}$

(b) $\dfrac{2}{3 - \sqrt{5}}$

Solution

(a) $\dfrac{3}{-\sqrt{5}} = \dfrac{3}{-\sqrt{5}} \cdot \dfrac{-\sqrt{5}}{-\sqrt{5}} = \dfrac{-3\sqrt{5}}{5}$

(b) $\dfrac{2}{3 - \sqrt{5}} = \dfrac{2}{3 - \sqrt{5}} \cdot \dfrac{3 + \sqrt{5}}{3 + \sqrt{5}} = \dfrac{2(3 + \sqrt{5})}{4} = \dfrac{3 + \sqrt{5}}{2}$

3.5 Solving equations involving surds

Solving equations involving surds

An equation that contains a surd or radical ($\sqrt{x}$) is called a radical equation. Often, solving a radical equation involves squaring a binomial. The binomial expansion formulas are shown below.

$$(x+y)^2 = x^2 + 2xy + y^2 \qquad \text{Common mistake: } (x+y)^2 \neq x^2 + y^2$$
$$(x-y)^2 = x^2 - 2xy + y^2 \qquad \text{Common mistake: } (x-y)^2 \neq x^2 - y^2$$

To solve a radical equation, square both sides of the equation to eliminate the square root. Then, solve for the variable. Once you get the solution of the radical equation, you need to substitute the solution in the original equation to check the solution. If the solution doesn't make the equation true, it is called an **extraneous solution** and is disregarded. Below shows how to solve the radical equation $x - 1 = \sqrt{x+5}$.

$$
\begin{array}{ll}
x - 1 = \sqrt{x+5} & \text{Square both sides} \\
(x-1)^2 = x+5 & \text{Use the binomial expansion formula} \\
x^2 - 2x + 1 = x + 5 & \text{Subtract } x+5 \text{ from each side} \\
x^2 - 3x - 4 = 0 & \text{Factor the quadratic expression} \\
(x+1)(x-4) = 0 & \text{Use the zero product property: If } ab = 0, \text{ then } a = 0 \text{ or } b = 0. \\
x = -1 \quad \text{or} \quad x = 4 &
\end{array}
$$

Substitute -1 and 4 for x in the original equation to check the solutions.

$$
\begin{array}{ll}
(-1) - 1 = \sqrt{-1+5} & (4) - 1 = \sqrt{4+5} \\
-2 \neq 2 \quad \text{(Not a solution)} & 3 = 3 \quad \checkmark \text{(Solution)}
\end{array}
$$

Example 7 Solving an equation involving surds

Solve the equation $\sqrt{3x-2} - \sqrt{2x-3} = 1$.

Solution

$$\sqrt{3x-2} - \sqrt{2x-3} = 1$$
$$\sqrt{3x-2} = 1 + \sqrt{2x-3}$$
$$(\sqrt{3x-2})^2 = (1 + \sqrt{2x-3})^2$$
$$3x - 2 = 1 + 2\sqrt{2x-3} + 2x - 3$$
$$x = 2\sqrt{2x-3}$$
$$x^2 = 4(2x-3)$$
$$x^2 - 8x + 12 = 0$$
$$(x-2)(x-6) = 0$$
$$x = 2 \quad \text{or} \quad x = 6$$

Substitute 2 and 6 for x in the original equation to check the solutions.

$$\sqrt{3(2)-2} - \sqrt{2(2)-3} = 1 \qquad\qquad \sqrt{3(6)-2} - \sqrt{2(6)-3} = 1$$
$$2 - 1 = 2 \quad \checkmark \text{ (Solution)} \qquad\qquad 4 - 3 = 1 \quad \checkmark \text{ (Solution)}$$

Therefore, the solutions to the equation $\sqrt{3x-2} - \sqrt{2x-3} = 1$ are $x = 2$ or $x = 6$.

EXERCISES

1. Solve the following equations.

    (a) $\sqrt{2x+3}+3=0$

    (b) $\sqrt{3x+2}-2\sqrt{x}=0$

    (c) $x-4=\sqrt{2x}$

    (d) $x+5=5-\sqrt{x}$

    (e) $\sqrt{x+3}-\sqrt{x-1}=1$

2. Solve the following simultaneous equations (system of equations).

$$\frac{9^x}{3^y} = 243$$
$$8^x \times 2^y = 1024$$

3. Rationalize the denominator and simplify.

(a) $\dfrac{1}{\sqrt{3} - 1}$

(b) $\dfrac{\sqrt{5} - \sqrt{3}}{\sqrt{5} + \sqrt{3}}$

(c) $\dfrac{1}{1 + \sqrt{2}} + \dfrac{1}{\sqrt{2} + \sqrt{3}} + \dfrac{1}{\sqrt{3} + 2}$

4. Find the value of k.

    (a) $10\sqrt{10}$ can be written in the form 10^k.

    (b) $\sqrt[4]{8 \times 2 \times 5^8} = k$.

    (c) $\dfrac{a^2\sqrt{a}}{\sqrt[4]{a^3}} = a^k$

5. Simplify the following expression.

$$(9 - 5\sqrt{32})(3 + 2\sqrt{2})$$

Give your answer in the form $a + b\sqrt{18}$.

6. Suppose $x = 2^a$ and $y = 2^b$. Find the values of a and b that satisfy the following simultaneous equations.

$$xy = 32$$
$$4xy^2 = 64$$

7. Given that a is a positive integer, show that

$$\sqrt{3a}\left(\sqrt{27a} + 2\sqrt{3a^5}\right)$$

is always a multiple of 3.

8. Answer the following questions. (Do not use a calculator in any part of the questions.)

   (a) Show that $2\sqrt{7} - 3\sqrt{2}$ is a square root of $46 - 12\sqrt{14}$.

   (b) Find the other root of $46 - 12\sqrt{14}$.

   (c) Express the expression $\dfrac{4\sqrt{7} + 5\sqrt{2}}{2\sqrt{7} - 3\sqrt{2}}$ in the form $\dfrac{a + b\sqrt{14}}{5}$, where a and b are integers.

CHAPTER 4 Factors of polynomials

4.1 Operations with polynomials

Polynomials

A polynomial is an expression of the form

$$a_n x^n + a_{n-1} x^{n-1} + a_{n-2} x^{n-2} + \cdots + a_2 x^2 + a_1 x + a_0$$

where:

- a_n is called the leading coefficient and $a_n \neq 0$

- All exponents (indices or powers) of the polynomial are whole numbers.

- The highest power of x in the polynomial, n, is called the **degree** of the polynomial.

- The coefficients a_n, a_{n-1}, $\cdots$, a_1, and a_0 are all real numbers.

- a_0 is called the constant term.

Below is a summary of common type of polynomials.

Degree	Type	Standard form
1	Linear	$a_1 x + a_0$
2	Quadratic	$a_2 x^2 + a_1 x + a_0$
3	Cubic	$a_3 x^3 + a_2 x^2 + a_1 x + a_0$
4	Quartic	$a_4 x^4 + a_3 x^3 + a_2 x^2 + a_1 x + a_0$

Operations on Polynomial Functions

Polynomial functions can be added, subtracted, and multiplied. Let $f(x) = x^2 + 1$ and $g(x) = x - 3$.

- The sum of $f(x)$ and $g(x)$: $\quad f(x) + g(x) = x^2 + 1 + x - 3 = x^2 + x - 2$

- The difference of $f(x)$ and $g(x)$: $\quad f(x) - g(x) = x^2 + 1 - (x - 3) = x^2 - x + 4$

- The product of $f(x)$ and $g(x)$: use the distributive property.

$$f(x) \cdot g(x) = (x^2 + 1)(x - 3) = x^3 - 3x^2 + x - 3$$

Polynomial Long Division and Synthetic Division

When dividing a polynomial by another polynomial, do polynomial long division or synthetic division. For instance, if you divide $x^3 - 2x^2 + 3x + 2$ by $x - 1$ doing polynomial long division or synthetic division as shown in Figure 1 and Figure 2, the quotient is $x^2 - x + 2$, and the remainder is 4.

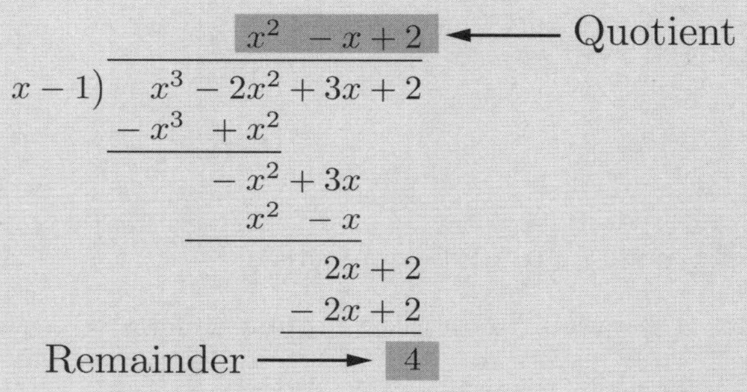

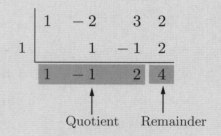

$$
\begin{array}{r}
\underline{x^2 \quad - x + 2} \longleftarrow \text{Quotient} \\
x - 1 \overline{) \; x^3 - 2x^2 + 3x + 2} \\
\underline{-x^3 + x^2} \\
-x^2 + 3x \\
\underline{x^2 - x} \\
2x + 2 \\
\underline{-2x + 2} \\
\text{Remainder} \longrightarrow \boxed{4}
\end{array}
$$

Fig. 1: Polynomial long division

Fig. 2: Synthetic division

Tip — If a polynomial $f(x)$ is divided by a divisor $d(x)$, the result can be written as follows:

- $f(x) = d(x) \cdot Q(x) + r \qquad$ or

- $\dfrac{f(x)}{d(x)} = Q(x) + \dfrac{r}{d(x)}$, where $Q(x)$ is the quotient polynomial and r is the remainder.

For instance, if $x^3 - 2x^2 + 3x + 2$ by $x - 1$, the result can be written as

- $x^3 - 2x^2 + 3x + 2 = (x - 1)(x^2 - x + 2) + 4 \qquad$ or

- $\dfrac{x^3 - 2x^2 + 3x + 2}{x - 1} = x^2 - x + 2 + \dfrac{4}{x - 1}$

4.2 Finding zeros of a polynomial function

Finding the Real Zeros of a Polynomial Function

A polynomial function $f(x)$ crosses the x-axis at three points as shown below.

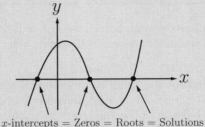

x-intercepts = Zeros = Roots = Solutions

These three points are called either the x-intercepts, zeros, roots, or solutions. That is, they are the values of x that make $f(x) = 0$.

Finding the real zeros of a polynomial function of degree 3 or higher is very complicated because there is no general formula like the quadratic formula available.

However, the remainder theorem, factor theorem, and rational zeros theorem are very useful to find the real zeros of a polynomial function.

4.3 The remainder theorem

> ### The Remainder Theorem
>
> - If a polynomial function $f(x)$ is divided by $x - k$, the remainder is $f(k)$.
> - If a polynomial function $f(x)$ is divided by $ax - b$, the remainder is $f\left(\dfrac{b}{a}\right)$.

 1. In order to evaluate the remainder, use either polynomial long division, synthetic division, or the remainder theorem. However, the remainder theorem is the easiest to use.

2. In order to determine the quotient polynomial when a polynomial function $f(x)$ is divided by $x - c$, use polynomial division or synthetic division. The remainder theorem doesn't tell you about the quotient polynomial.

	Quotient	Remainder
Polynomial long division	✓	✓
Synthetic division	✓	✓
Remainder theorem	Not available	✓

Example 1 Finding the remainder

If $f(x) = x^{100} - x^{99} + 5x + 1$ is divided by $x + 1$, find the remainder using the following.

(a) the remainder theorem

(b) polynomial long division

Solution

(a) Solving $x + 1 = 0$ gives $x = -1$. Thus, the remainder, $r = f(-1)$, is as follows.

$$
\begin{aligned}
r &= f(-1) \\
&= (-1)^{100} - (-1)^{99} + 5(-1) + 1 \\
&= 1 + 1 - 5 + 1 \\
&= -2
\end{aligned}
$$

(b) In order to find the remainder, you need to divide $x^{100} - x^{99} + 5x + 1$ by $x + 1$ 100 times, which is a very tedious procedure. The polynomial long division is not recommended when the degree of a polynomial is very large.

Example 2   Using the remainder theorem

Let $f(x) = 4x^3 + ax^2 - 11x + b$.

    When $f(x)$ is divided by $x - 1$, the remainder is -9.

    When $f(x)$ is divided by $x + 2$, the remainder is -33.

Find the values of a and b.

Solution $f(x) = 4x^3 + ax^2 - 11x + b$.

- When $f(x)$ is divided by $x - 1$, the remainder is -9, which means $r = f(1) = -9$. Thus,

$$f(1) = -9$$
$$4 + a - 11 + b = -9$$
$$a + b = -2$$

- When $f(x)$ is divided by $x + 2$, the remainder is -33, which means $r = f(-2) = -33$. Thus,

$$f(-2) = -33$$
$$-32 + 4a + 22 + b = -33$$
$$4a + b = -23$$

Solving the simultaneous equations $a + b = -2$ and $4a + b = -23$ gives $a = -7$ and $b = 5$.

4.4 The factor theorem

The Factor Theorem

- For a polynomial function $f(x)$, if $f(k) = 0$, then $x - k$ is a factor of $f(x)$
- For a polynomial function $f(x)$, if $f\left(\dfrac{b}{a}\right) = 0$, then $ax - b$ is a factor of $f(x)$.

Tip   The factor theorem states a relationship between a zero and a factor such that if a function has a zero k, then the function has a factor of $x - k$.

Example 3   Finding a factor of a polynomial function

Determine whether the function $f(x) = x^3 - 7x + 6$ has the factor $x - 2$.

Solution   Use the factor theorem to determine whether $x - 2$ is the factor of $f(x)$. Substitute 2 for x in $f(x)$ to see if the remainder $r = f(2)$ is equal to 0.

$$f(x) = x^3 - 7x + 6 \qquad\qquad \text{Substitute 2 for } x$$
$$f(2) = 2^3 - 7(2) + 6 = 0$$

Since the remainder $r = f(2) = 0$, $x - 2$ is a factor of $f(x) = x^3 - 7x + 6$.

Example 4   Using the factor theorem

If $2x^2 - 3x + 1$ is a factor of $2x^3 - 9x^2 + ax + b$, find the values of a and b.

Solution   Let $f(x) = 2x^3 - 9x^2 + ax + b$.

Since $2x^2 - 3x + 1 = (2x - 1)(x - 1)$ is a factor of $2x^3 - 9x^2 + ax + b$, $2x - 1$ and $x - 1$ are also factors of $f(x)$. Using the factor theorem, $f\left(\dfrac{1}{2}\right) = 0$ and $f(1) = 0$.

- When $f\left(\dfrac{1}{2}\right) = 0$:

$$f\left(\frac{1}{2}\right) = 0$$
$$\frac{1}{4} - \frac{9}{4} + \frac{a}{2} + b = 0$$
$$\frac{a}{2} + b = 2$$
$$a + 2b = 4$$

- When $f(1) = 0$:

$$f(1) = 0$$
$$2 - 9 + a + b = 0$$
$$a + b = 7$$

Solving the simultaneous equations $a + 2b = 4$ and $a + b = 7$ gives $a = 10$ and $b = -3$.

4.5 Rational zeros theorem

3. Rational Zeros Theorem

The rational zeros theorem provides a list of all possible rational zeros of a polynomial function with integer coefficients. Out of all possible rational zeros, use the factor theorem or synthetic division to find the real zeros of the polynomial function.

The rational zeros theorem states that for the polynomial function with integer coefficients, $f(x) = a_n x^n + a_{n-1} x^{n-1} + \cdots + a_1 x + a_0$, the possible rational zeros are as follows:

$$\text{Possible rational zeros} = \pm \frac{\text{factors of } a_o}{\text{factors of } a_n}$$

where a_n is the leading coefficient and a_0 is the constant.

For instance, if $f(x) = 2x^3 - 11x^2 + 13x - 4$, the possible rational zeros are shown below.

$$\text{Possible rational zeros} = \pm \frac{\text{factors of } 4}{\text{factors of } 2}$$

$$= \pm \frac{\{1, 2, 4\}}{\{1, 2\}}$$

$$= \pm 1, \ \pm 2, \ \pm 4, \ \pm \frac{1}{2}$$

There are 8 possible rational zeros. Use the factor theorem to find one real zero of $f(x)$.

$$f(-1) = -30 \neq 0, \qquad f(1) = 0$$

$f(1) = 0$ means that 1 is a zero of $f(x)$. Let's use synthetic division to factor $f(x)$.

$$
\begin{array}{r|rrrr}
 & 2 & -11 & 13 & -4 \\
1 & & 2 & -9 & 4 \\
\hline
 & 2 & -9 & 4 & 0
\end{array}
$$

Since the quotient polynomial is $2x^2 - 9x + 4$, the function $f(x)$ can be factored as follows:

$$2x^3 - 11x^2 + 13x - 4 = (x - 1)(2x^2 - 9x + 4) \qquad \text{Factor } 2x^2 - 9x + 4$$

$$= (x - 1)(x - 4)(2x - 1)$$

Therefore, the real zeros of $f(x)$ are 1, 4, and $\frac{1}{2}$.

Tip   If a polynomial function has a rational zero, the zero is one of the possible rational zeros suggested by the rational zeros theorem. However, if a polynomial function does not have any rational zeros, the rational zeros theorem does not help you find complex zeros.

Conjugate Pairs Theorem

The conjugate pairs theorem states that complex zeros and irrational zeros always occur in conjugate pairs.

$$\text{If } a + bi \text{ is a zero of } f, \qquad \Longrightarrow \qquad a - bi \text{ is also a zero of } f.$$
$$\text{If } a + \sqrt{b} \text{ is a zero of } f, \qquad \Longrightarrow \qquad a - \sqrt{b} \text{ is also a zero of } f.$$

Writing a Polynomial Function with the Given Zeros

By the factor theorem, if 2 is a zero of f, then $x - 2$ is a factor of f. In order to write a polynomial function with the given zeros, convert each zero to a factor and expand. For instance,

$$\text{Given zeros: 2, and 3} \quad \Longrightarrow \quad y = (x-2)(x-3) = x^2 - 5x + 6$$

$$
\begin{aligned}
\text{Given zeros: } 1 + \sqrt{2}, \text{ and } 1 - \sqrt{2} \quad \Longrightarrow \quad y &= (x - (1 + \sqrt{2}))(x - (1 - \sqrt{2})) \\
&= ((x-1) - \sqrt{2})((x-1) + \sqrt{2}) \\
&= (x-1)^2 - 2 \\
&= x^2 - 2x - 1
\end{aligned}
$$

Vieta's Formulas

When irrational zeros or complex zeros are given, use Vieta's formula to easily write a quadratic function with leading coefficient 1. Vieta's formulas relate the coefficients of a polynomial to the sum and product of its zeros and are described below. For a quadratic function $f(x) = x^2 + bx + c$, let z_1 and z_2 be the zeros of f.

$$z_1 + z_2 = -b \qquad \text{Sum of zeros equals the opposite of the coefficient of } x$$
$$z_1 z_2 = c \qquad \text{Product of zeros equals the constant term}$$

For instance, when irrational zeros, $1 + \sqrt{2}$ and $1 - \sqrt{2}$, are given,

$$\text{Sum of zeros:} \quad (1 + \sqrt{2}) + (1 - \sqrt{2}) = 2 \quad \xrightarrow{\text{Opposite}} \quad -2 \text{ (Coefficient of } x)$$
$$\text{Product of zeros:} \quad (1 + \sqrt{2})(1 - \sqrt{2}) = -1 \quad \xrightarrow{\text{Same}} \quad -1 \text{ (Constant term)}$$

Thus, the quadratic function whose zeros are $1 + \sqrt{2}$ and $1 - \sqrt{2}$ is $x^2 - 2x - 1$.

1. In order to expand the expression $((x-1) - \sqrt{2})((x-1) + \sqrt{2})$ shown above, use $(a-b)(a+b) = a^2 - b^2$ formula.

2. In order to use Vieta's formulas, two zeros must be the same type: rational zeros, irrational zeros, or complex zeros.

Example 5 Writing a polynomial function with given zeros

If zeros of the polynomial function f are 1, and $2 - 3\sqrt{3}$, write a cubic function with leading coefficient of 1.

Solution   According to the conjugate pairs theorem, $2 + 3\sqrt{3}$, is also a zero of f. Thus, 1, $2 - 3\sqrt{3}$, and $2 + 3\sqrt{3}$ are zeros of f. Since irrational zeros are given, use Vieta's formulas to write a quadratic function with leading coefficient of 1.

$$\text{Sum of zeros:}\quad (2 + 3\sqrt{3}) + (2 - 3\sqrt{3}) = 4 \quad \xrightarrow{\text{Opposite}} \quad -4 \text{ (Coefficient of } x)$$

$$\text{Product of zeros:}\quad (2 + 3\sqrt{3})(2 - 3\sqrt{3}) = -23 \quad \xrightarrow{\text{Same}} \quad -23 \text{ (Constant term)}$$

The quadratic function whose zeros are $2 + 3\sqrt{3}$ and $2 - 3\sqrt{3}$ is $x^2 - 4x - 23$. Since 1 is a zero of f, $x - 1$ is a factor of f. Thus,

$$\begin{aligned} f(x) &= (x - 1)(x^2 - 4x - 23) \\ &= x^3 - 5x^2 - 19x + 23 \end{aligned}$$

Therefore, the cubic function with leading coefficient of 1 is $f(x) = x^3 - 5x^2 - 19x + 23$.

4.6 Graphing cubic functions

Graphing Cubic Functions

A polynomial function of degree 3, $f(x) = ax^3 + bx^2 + cx + d$, is a cubic function, where a is the leading coefficient and is nonzero. Depending on the values of a, either positive or negative, the graph of the cubic function varies.

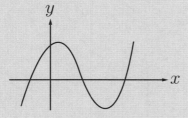

Figure 1: If $a > 0$

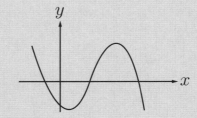

Figure 2: If $a < 0$

If $a > 0$, the graph of a cubic function shown in Figure 1 goes up as x increases and goes down as x decreases. Whereas, if $a < 0$, the graph of a cubic function shown in Figure 2 goes down as x increases and goes up as x decreases.

Zeros of odd or even multiplicity

If $(x - c)^m$ is a factor of a polynomial function f, c is called a zero of multiplicity of m. Depending on the value of m, either odd or even, the graph of f either crosses or touches the x-axis at $x = c$.

If $m = $ odd $\implies$ graph of f crosses the x-axis at $x = c$.

If $m = $ even $\implies$ graph of f touches the x-axis at $x = c$.

For instance, let $f(x) = -5(x - 1)^2(x - 3)$. Since the leading coefficient is -5, the shape of the graph of f is similar to the graph in Figure 2.

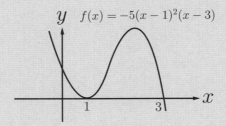

Figure 3

Since $(x - 1)^2$ is a factor of f, 1 is a zero of multiplicity 2. Thus, the graph of f touches the x-axis at $x = 1$. Additionally, $(x - 3)$ is a factor of f. Thus, 3 is a zero of multiplicity 1 and the graph of f crosses the x-axis at $x = 3$. The graph of f is shown in Figure 3 above.

4.7 Solving cubic inequalities graphically

Solving Polynomial Inequalities

Solving a polynomial inequality means finding the x-values for which the graph of the polynomial function f lies above or below the x-axis. A polynomial inequality can be solved algebraically. However, solving a polynomial inequality **graphically** is highly recommended.

Let the polynomial inequality be $(x-1)(x-2)(x-3) > 0$. First, graph the polynomial function $f(x) = (x-1)(x-2)(x-3)$ with the leading coefficient 1. Since 1, 2, and 3 are zeros of multiplicity 1, the graph of f crosses the x-axis at $x = 1$, $x = 2$, and $x = 3$ as shown in Figure 4.

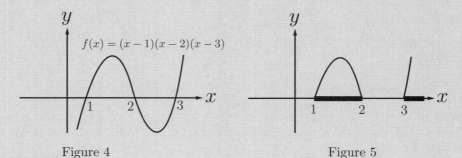

Figure 4   Figure 5

In Figure 5, the graph of f lies above the x-axis when $1 < x < 2$ or $x > 3$. Therefore, the solution to $(x-1)(x-2)(x-3) > 0$ is $1 < x < 2$ or $x > 3$.

EXERCISES

1. Divide using polynomial long division.

   (a) $\dfrac{x^3 - 3x^2 + x - 9}{x - 1}$

   (b) $\dfrac{2x^3 + 5x^2 - 3x + 6}{x + 2}$

   (c) $\dfrac{10x^3 + 27x^2 + 14x + 7}{x^2 + 2}$

2. If $(x-a)$ is a factor of $2x^3 + (a-1)x^2 - (a+2)x + a$. Find the all possible values of a.

3. Write the cubic function with the leading coefficient of 2 if the roots are $\frac{1}{2}$, $1 + \sqrt{3}$.

4. Let $f(x) = x^3 + ax^2 - 14x + b$. $f(x)$ has a factor of $x + 2$ and leaves a remainder of -10 when divided by $x + 1$. Find the values of a and b.

5. Find the all zeros of the function $f(x) = 2x^3 - 5x^2 - 14x + 8$.

6. The polynomial $2x^3 + 4x^2 - 3x - 9$ can be written as

$$2x^3 + 4x^2 - 3x - 9 = ax(x+1)^2 + b(x+1) + C$$

Find the values of a, b, and c.

7. Factorize $x^3 - 6x^2 - x + 30$ completely .

8. Solve the following cubic inequalities.

    (a) $(x+1)(x-2)(x-3) \leq 0$

    (b) $-2(x-1)(x+3)(x+2) > 0$

9. Let $f(x) = ax^3 - 4x^2 + bx + 3$. If $f(x)$ is divided by $x-1$, the remainder is -2. If $f(x)$ is divided by $x-2$, the remainder is 9.

    (a) Find the values of a and b.

    (b) Find the remainder when $f(x)$ is divided by $x+2$.

10. $f(x) = x^3 - 7x^2 + 2x + 40$.

    (a) Factorize $f(x)$ completely.

    (b) Solve the inequality $x^3 - 7x^2 + 2x + 40 \leq 0$.

11. The polynomial $f(x) = 6x^3 + ax^2 + bx + 1$ is divisible by $2x - 1$. The remainder when $f(x)$ is divided by $x + 2$ is 39 less than the remainder when $f(x)$ is divided by $x - 1$.

   (a) Show that $a = 1$ and find the value of b.

   (b) Show that $f(x)$ can be written as $f(x) = (3x - 1)(cx^2 + dx + e)$, where c, d, and e are integers.

   (c) Factorize $f(x)$ completely.

   (d) Solve $f(x) = 0$.

   (e) Solve $f(x) > 0$.

CHAPTER 5 Logarithmic and exponential functions

5.1 Logarithms

Definition of Logarithm

We know that $10^1 = 10$ and $10^2 = 100$. For what value of x satisfy $10^x = 45$? Since $10^1 < 10^x < 10^2$, we expect x to be between 1 and 2. To find the exact value of x, mathematicians defined **logarithms**. In terms of a logarithm, $x = \log_{10} 45 \approx 1.653$.

In general,

$$a^x = y \qquad \Longleftrightarrow \qquad x = \log_a y$$

- Two equations $a^x = y$ and $x = \log_a y$ are equivalent.

- Exponential form: $a^x = y$

- Logarithmic form: $x = \log_a y$

- The expression $\log_a y$ is read as 'log base a of y'.

- $\log_{10} x$ can also be written as $\log x$ or $\lg x$.

Example 1   Converting to a logarithmic and exponential form

Convert the following equations.

(a) Convert $\log_5 x = 3$ to exponential form.

(b) Convert $3^x = 81$ to logarithmic form

Solution

(a) $\log_5 x = 3 \quad \Longleftrightarrow \quad x = 5^3.$

(b) $3^x = 81 \quad \Longleftrightarrow \quad x = \log_3 81.$

Example 2 Evaluating logarithmic expressions

Without using a calculator, evaluate the following expressions.

(a) $\lg 1000$

(b) $\log_2 \sqrt{2}$

(c) $\log_5 \dfrac{1}{125}$

Solution

(a) Let $x = \lg 1000$ and then convert $x = \lg 1000$ to exponential form.

$$x = \lg 1000$$
$$x = \log_{10} 10^3$$
$$10^x = 10^3$$
$$x = 3$$

Thus, $x = \lg 1000 = 3$.

(b) Let $x = \log_2 \sqrt{2}$ and then convert $x = \log_2 \sqrt{2}$ to exponential form.

$$x = \log_2 \sqrt{2}$$
$$2^x = \sqrt{2}$$
$$2^x = 2^{\frac{1}{2}}$$
$$x = \frac{1}{2}$$

Thus, $x = \log_2 \sqrt{2} = \dfrac{1}{2}$.

(c) Let $x = \log_5 \dfrac{1}{125}$ and then convert $x = \log_5 \dfrac{1}{125}$ to exponential form.

$$x = \log_5 \frac{1}{125}$$
$$5^x = 5^{-3}$$
$$x = -3$$

Thus, $x = \log_5 \dfrac{1}{125} = -3$.

5.2 Properties of logarithms

Properties of Logarithms

The table below summarizes the properties of logarithms.

Properties of Logarithms	Example
1. $\log_a 0 = $ undefined	1. $\log_a 0 = $ undefined
2. $\log_a 1 = 0$	2. $\log_2 1 = 0$
3. $\log_a a = 1$	3. $\log_2 2 = 1$
4. $\log_a x^n = n \log_a x$	4. $\log_2 125 = \log_2 5^3 = 3 \log_2 5$
5. $\log_{a^n} x = \frac{1}{n} \log_a x$	5. $\log_8 5 = \log_{2^3} 5 = \frac{1}{3} \log_2 5$
6. $\log_a x = \frac{\log_c x}{\log_c a}$	6. $\log_2 3 = \frac{\log_{10} 3}{\log_{10} 2} = \frac{\ln 3}{\ln 2}$
7. $\log_a x = \frac{1}{\log_x a}$	7. $\log_2 3 = \frac{1}{\log_3 2}$
8. $\log_a xy = \log_a x + \log_a y$	8. $\log_2 15 = \log_2(3 \cdot 5) = \log_2 3 + \log_2 5$
9. $\log_a \frac{x}{y} = \log_a x - \log_a y$	9. $\log_2 \frac{5}{3} = \log_2 5 - \log_2 3$
10. $a^{\log_a x} = x^{\log_a a} = x$	10. $2^{\log_2 3} = 3^{\log_2 2} = 3^1 = 3$

 1. The common logarithm is the logarithm with base 10, which is denoted by either $\log_{10} x$, $\log x$, or $\lg x$.

2. e is an irrational number and is approximately $2.718\cdots$. The natural logarithm of x can be expressed as either $\log_e x$ or $\ln x$.

Example 3   Writing a logarithmic expression

If $\log 2 = x$ and $\log 3 = y$, write $\log 72$ in terms of x and y.

Solution   Since the prime factorization of $72 = 2^3 \cdot 3^2$,

$$\begin{aligned}
\log 72 &= \log(2^3 \cdot 3^2) &&\text{Use } \log_a xy = \log_a x + \log_a y \\
&= \log 2^3 + \log 3^2 &&\text{Use } \log_a x^n = n \log_a x \\
&= 3 \log 2 + 2 \log 3 &&\text{Since } \log 2 = x \text{ and } \log 3 = y \\
&= 3x + 2y
\end{aligned}$$

Therefore, $\log 72$ can be written as $3x + 2y$.

Example 4   Writing the expression as a single logarithm

Write the following expression as a single logarithm.

(a) $2\log_5 x + 3\log_5 x + 4\log_5 y$

(b) $3(\ln 3 - \ln x) + (\ln x - \ln 9)$

Solution

(a)

$$\begin{aligned}
2\log_5 x + 3\log_5 x + 4\log_5 y &= \log_5 x^2 + \log_5 x^3 + \log_5 y^4 \\
&= \log_5 x^2 x^3 y^4 \\
&= \log_5 x^5 y^4
\end{aligned}$$

(b)

$$\begin{aligned}
3(\ln 3 - \ln x) + (\ln x - \ln 9) &= 3\ln\frac{3}{x} + \ln\frac{x}{9} \\
&= \ln\left(\frac{3}{x}\right)^3 + \ln\frac{x}{9} \\
&= \ln\frac{27}{x^3} + \ln\frac{x}{9} \\
&= \ln\left(\frac{27}{x^3}\cdot\frac{x}{9}\right) \\
&= \ln\frac{3}{x^2}
\end{aligned}$$

Example 5   Proving change of base of logarithms

Prove that $\log_a b = \dfrac{\log_c b}{\log_c a}$.

Solution   Let $x = \log_a b$ and convert it to exponential form.

$$\begin{aligned}
x &= \log_a b \\
a^x &= b \\
\log_c a^x &= \log_c b \\
x\log_c a &= \log_c b \\
x &= \frac{\log_c b}{\log_c a}
\end{aligned}$$

Example 6   Simplifying the expression using change of base of logarithms

Show that $\log_4(x-2) = \dfrac{1}{2}\log_2(x-2)$ using change of base of logarithms.

Solution

$$\log_4(x-2) = \frac{\log_2(x-2)}{\log_2 4}$$
$$= \frac{\log_2(x-2)}{\log_2 2^2}$$
$$= \frac{\log_2(x-2)}{2\log_2 2}$$
$$= \frac{\log_2(x-2)}{2}$$
$$= \frac{1}{2}\log_2(x-2)$$

Tip

Using the property of logarithm $\log_{a^n} x = \frac{1}{n}\log_a x$, it is easy to show that $\log_4(x-2) = \dfrac{1}{2}\log_2(x-2)$.

$$\log_4(x-2) = \log_{2^2}(x-2) = \frac{1}{2}\log_2(x-2)$$

5.3 Solving exponential and logarithmic Equations

Solving Exponential and Logarithmic Equations

Below shows how to solve an exponential equation and a logarithmic equation.

- When each side of an equation (either exponential or logarithmic) has the same base.

$$\text{If } a^x = a^y \implies x = y \qquad\qquad \text{e.g. If } 2^x = 2^3, \text{ then } x = 3$$
$$\text{If } \log_a x = \log_a y \implies x = y \qquad \text{e.g. If } \log_2 x = \log_2 5, \text{ then } x = 5$$

- When each side of an equation (either exponential or logarithmic) has a different base, convert the exponential equation to a logarithmic equation or vice versa.

$$\text{If } a^x = b \implies x = \log_a b \qquad\qquad \text{e.g. If } 2^x = 7, \text{ then } x = \log_2 7$$
$$\text{If } \log_a x = b \implies x = a^b \qquad\qquad \text{e.g. If } \log_2 x = 3, \text{ then } x = 2^3$$

Tip   Always check you solutions whenever you solve the following types of equations because some solutions may be extraneous.

- Absolute value equations: e.g. $|x + 6| = 2x$

- Radical equations: e.g. $x - 1 = \sqrt{x + 5}$

- Rational equations: e.g. $\dfrac{1}{x - 1} = \dfrac{2}{x(x - 1)}$

- Logarithmic equations: e.g. $\ln(x + 1) + \ln(x - 2) = \ln 4$

Example 7   Solving an exponential equation

If $2^x = 3^y$, find the value of $\frac{x}{y}$.

Solution   Since each side of equation has a different base (left side has the base 2 and right side has the base 3), convert the exponential equation to a logarithmic equation.

$$
\begin{aligned}
2^x &= 3^y & &\text{Convert the equation to a logarithmic equation} \\
x &= \log_2 3^y & &\text{Use } \log_a x^n = n \log_a x \\
x &= y \log_2 3 & &\text{Divide each side by } y \\
\frac{x}{y} &= \log_2 3
\end{aligned}
$$

Therefore, the value of $\frac{x}{y}$ is $\log_2 3 \approx 1.585$.

Example 8   Solving a logarithmic equation

Solve $\ln(x+1) + \ln(x-2) = \ln 4$.

Solution   Express the left side as a single logarithm and solve for x.

$$\ln(x+1) + \ln(x-2) = \ln 4 \qquad \text{Express the left side as a single logarithm}$$
$$\ln\left[(x+1)(x-2)\right] = \ln 4 \qquad \text{Since each side has the same base, } e$$
$$(x+1)(x-2) = 4 \qquad \text{Subtract 4 from each side}$$
$$x^2 - x - 6 = 0 \qquad \text{Use the factoring method}$$
$$(x+2)(x-3) = 0 \qquad \text{Use the zero product property}$$
$$x = -2 \quad \text{or} \quad x = 3$$

Substitute -2 and 3 for x in the original equation to check the solutions.

$$\ln(-2+1) + \ln(-2-2) = \ln 4 \qquad\qquad \ln(3+1) + \ln(3-2) = \ln 4$$
$$\text{undefined} \neq \ln 4 \quad \text{(Not a solution)} \qquad\qquad \ln 4 = \ln 4 \quad \checkmark \text{ (Solution)}$$

Therefore, the only solution to $\ln(x+1) + \ln(x-2) = \ln 4$ is $x = 3$.

Example 9   Solving a logarithmic equation

Solve $6 \log_x 3 - \log_x 9 = 4$.

Solution

$$6 \log_x 3 - \log_x 9 = 4$$
$$6 \log_x 3 - \log_x 3^2 = 4$$
$$6 \log_x 3 - 2 \log_x 3 = 4 \qquad\qquad \text{Factor } \log_x 3$$
$$\log_x 3(6-2) = 4$$
$$4 \log_x 3 = 4$$
$$\log_x 3^4 = 4$$
$$x^4 = 3^4$$
$$x = \pm 3$$

Since logarithms only exist for positive bases, $x = -3$ is not a solution. Substitute 3 for x in the original equation to check the solutions.

$$6 \log_3 3 - \log_3 3^2 = 6 - 2 \log_3 3 = 6 - 2 = 4$$

So $x = 3$ satisfies the original equation. Therefore, the solution to the equation is $x = 3$.

5.4 Graphs of logarithmic and exponential functions

Graphs of Logarithmic and Exponential Functions

The exponential functions, $y = a^x$, and logarithmic functions, $y = \log_a x$, are inverses of each other. For instance, the inverse function of 3^x is $\log_3 x$. The following two equations relate exponents to logarithms or vice versa and are used to find the inverse function of an exponential function or a logarithmic function.

$$a^x = y \qquad \Longleftrightarrow \qquad x = \log_a y$$

Note that the two equations above are equivalent.

Since exponential functions and logarithmic functions are inverse functions, their graphs are symmetric with respect to the line $y = x$ as shown in Figures 1 and 2.

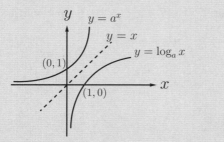

Figure 1: when $a > 1$

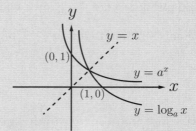

Figure 2: when $0 < a < 1$

Tip — The graphs of all exponential functions pass through the point $(0, 1)$. Whereas, the graphs of all logarithmic functions pass through the point $(1, 0)$.

Domain of a Logarithmic function

For a logarithmic function $y = \log_a h(x)$, where $h(x)$ is an algebraic expression or the argument of a logarithmic function, the domain of a logarithmic function is a set of all x-values for which $h(x) > 0$. In other words, solve the inequality $h(x) > 0$ to find the domain of a logarithmic function. For instance,

$$\log_2(x - 2) \implies \text{Solve } x - 2 > 0 \implies \text{Domain: } x > 2$$
$$\log_3(6 - 2x) \implies \text{Solve } 6 - 2x > 0 \implies \text{Domain: } x < 3$$
$$\log_{10}(x^2 - 2x) \implies \text{Solve } x^2 - 2x > 0 \implies \text{Domain: } x < 0 \text{ or } x > 2$$

5.5 Graphs of $y = ke^{nx} + a$ and $y = k\ln(ax+b)$

Graphs of $y = ke^{nx} + a$

The graphs of the exponential functions $y = ke^{nx} + a$ have the following characteristics.

- y-intercept (when $x = 0$) is $k + a$.
- The horizontal asymptote is $y = a$.
- When $k > 0$ and $n > 0$: As $x \to \infty$, $y \to \infty$.
- When $k < 0$ and $n > 0$: As $x \to \infty$, $y \to -\infty$.
- When $k > 0$ and $n < 0$: As $x \to \infty$, $y \to a$.
- When $k < 0$ and $n < 0$: As $x \to \infty$, $y \to a$.

Graphs of $y = k\ln(ax + b)$

The graphs of the logarithmic functions $y = k\ln(ax + b)$ have the following characteristics.

- The domain (solve for $ax + b > 0$) is $x > -\dfrac{b}{a}$.
- The vertical asymptote is $x = -\dfrac{b}{a}$.
- When $k > 0$, as $x \to \infty$, $y \to \infty$.
- When $k < 0$, as $x \to \infty$, $y \to -\infty$.

Example 10   Graphing a exponential function $y = ke^{nx} + a$

Sketch the graph of $y = 2e^{-3x} - 4$.

Solution   The graphs of the exponential function $y = 2e^{-3x} - 4$ have the following characteristics.

- y-intercept (when $x = 0$) is -2.

- The horizontal asymptote is $y = -4$.

- For $n = -3 < 0$, as $x \to \infty$, $y \to -4$.

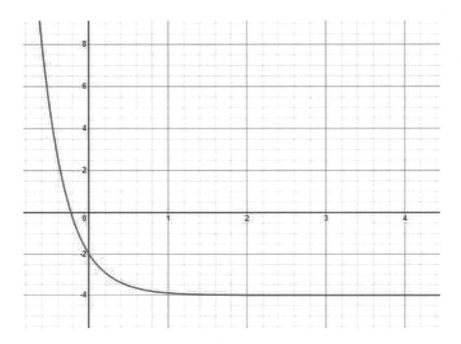

Example 11   **Graphing a logarithmic function** $y = k\ln(ax + b)$

Sketch the graph of $y = 2\ln(3x + 6)$

Solution   The graphs of the logarithmic functions $y = 2\ln(3x + 6)$ have the following characteristics.

- The domain (solve for $3x + 6 > 0$) is $x > -2$.

- The vertical asymptote is $x = -2$.

- For $k = 2 > 0$, as $x \to \infty$, $y \to \infty$.

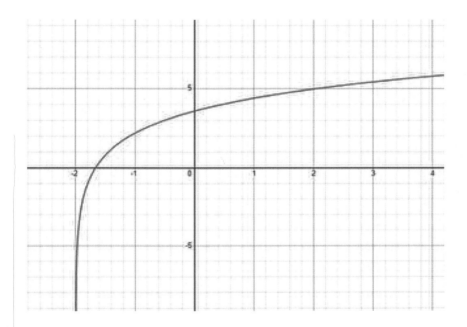

EXERCISES

1. Sketch the following exponential functions.

    (a) $y = 4e^{2x} + 5$

    (b) $y = 3e^{-4x} - 1$

    (c) $y = 2 - 2e^{3x}$

    (d) $y = -1 - e^{-x}$

2. Sketch the following logarithmic functions.

    (a) $y = -2\ln(2x - 4)$

    (b) $y = 3\ln(3x - 6)$

    (c) $y = 4\ln(x + 2)$

    (d) $y = -2\ln(3x + 6)$

3. Using the substitution $y = 3^x$,

    (a) Show that the equation $3^{2x+1} - 3^{x+1} + 2 = 4(3^x)$ can be written in the form $ay^2 + by + 2 = 0$, where a and b are constants to be found.

    (b) Solve the equation $3^{2x+1} - 3^{x+1} + 2 = 4(3^x)$.

4. Solve the following simultaneous equations.

$$\ln x = 3 \ln y$$
$$3^x = 27^y$$

5. Write each expression as a single logarithm.

   (a) $3 \log x + 4 \log y$

   (b) $\log_2 \left(\dfrac{x}{x-1} \right) + \log_2 \left(\dfrac{x+1}{x} \right) - \log_2 (x^2 - 1)$

   (c) $\dfrac{1}{2} \ln(x^2 + 1) - \left[\dfrac{1}{2} \ln(x - 2) + \ln x \right]$

6. Consider the function $f(x) = -\dfrac{1}{2} \ln(2x + 3)$.

   (a) Find the domain and range of f.

   (b) Find the vertical asymptote if any.

   (c) Find the inverse function f^{-1}.

   (d) Find the domain and range of f^{-1}.

7. Solve the following equations.

    (a) $\lg x + \lg(x - 2) = \lg(x + 4)$

    (b) $\log_5 x = \log_{25}(2x + 8)$

    (c) $5^{x-2} = 3^{3x+2}$

8. If $\log_a x = 2$, and $\log_a y = 3$, evaluate the following expressions.

    (a) $\log_a x^2 y$

    (b) $\log_{x^2} a$

    (c) $\log_{xy} \sqrt{a}$

9. The number N of bacteria present in a culture at time t (in minutes) obeys the function $N(t) = 100e^{0.5t}$ for $t \geq 0$.

   (a) Find the initial population.

   (b) After how many minutes will the population equal 1000 ?

10. The temperature, $T°$ Celsius, of an object, t minutes after it is removed from a heat source, is given by

$$T(t) = 69e^{kt} + 24, \qquad \text{where } k < 0 \text{ and } t \geq 0$$

   (a) Sketch the graph of $T(t)$.

   (b) Find the temperature of the object at the instance it is removed from the heat source.

   (c) If the temperature of the object is $75°$ C after 10 minutes, find the value of k.

   (d) Find the temperature of the object when $t = 15$ minutes.

11. Answer the following questions. (Do not use a calculator in this question.)

   (a) Simplify the following expression

   $$3^{3\log_3 x}$$

   in terms of x.

   (b) Find the range of k for which the equation $e^{-2x} - 1 = k$ has no solutions.

   (c) Solve the following simultaneous equations.

   $$9^{x+2} \times 81^{y-1} = 243$$
   $$8^{x-2} \times 2^{y+8} = 128$$

 **CHAPTER 6** | Straight-line graphs

6.1 Coordinate geometry

Coordinate Geometry

The **gradient** or **slope**, m, of a line is a number that describes the steepness of the line. The larger the absolute value of the gradient, $|m|$, the steeper the line is (closer to y-axis). If a line passes through the points (x_1, y_1) and (x_2, y_2), the gradient m is defined as

$$m = \frac{\text{Rise}}{\text{Run}} = \frac{y_2 - y_1}{x_2 - x_1}$$

If the points $A(x_1, y_1)$ and $B(x_2, y_2)$ are given, the following formulas are useful in solving math problems.

Midpoint Formula: $\left(\dfrac{x_1 + x_2}{2}, \dfrac{y_1 + y_2}{2} \right)$

Distance Formula: $AB = \sqrt{(x_2 - x_1)^2 + (y_2 - y_1)^2}$

An equation of a line can be written in three different forms.

1. **Slope-intercept form:** $y = mx + b$, where m is slope and b is y-intercept.

2. **Point-slope form:** If the slope of a line is m and the line passes through the point (x_1, y_1),

$$y - y_1 = m(x - x_1)$$

3. **Standard form:** $Ax + By = C$, where A, B, and C are integers.

Below summarizes the lines by gradient.

- Lines that rise from left to right have positive gradient.

- Lines that fall from left to right have negative gradient.

- Horizontal lines have zero gradient (example: $y = 2$).

- Vertical lines have undefined gradient (example: $x = 2$).

- **Parallel** lines have the same gradient.

- **Perpendicular** lines have negative reciprocal gradient (product of the gradients equals -1).

- If three points A, B, C are **collinear** (they lie on the same line),

$$\text{Gradient of } AB = \text{Gradient of } BC$$

The x-intercept of a line is a point where the line crosses x-axis. **The y-intercept** of a line is a point where the line crosses y-axis.

To find the x-intercept of a line	$\Longrightarrow$	Substitute 0 for y and solve for x
To find the y-intercept of a line	$\Longrightarrow$	Substitute 0 for x and solve for y

Example 1   Finding the perpendicular bisector of a segment

Find the equation of the perpendicular bisector of segment $\overline{AB}$ whose endpoints are $A(3, 3)$ and $B(5, -1)$.

Solution   The midpoint of the segment $\overline{AB}$ is $\left(\dfrac{3+5}{2}, \dfrac{3+-1}{2} \right) = (4, 1)$. The gradient of the segment AB is $\dfrac{-1-3}{5-3} = -2$. The gradient of the perpendicular bisector is the negative reciprocal of -2, which is $\dfrac{1}{2}$. Using the point-slope form of a line with $m = \dfrac{1}{2}$ and the midpoint $(4, 1)$,

$$y - 1 = \frac{1}{2}(x - 4) \qquad \text{Point-slope form}$$
$$y = \frac{1}{2}x - 1$$

Therefore, the equation of the perpendicular bisector of $\overline{AB}$ is $y = \dfrac{1}{2}x - 1$.

Example 2   Finding the distance between two points

The distance between two points $C(a, -3)$ and $D(3, a - 2)$ is $4\sqrt{5}$. Find the two possible values of a.

Solution   Using the distance formula $CD = \sqrt{(x_2 - x_1)^2 + (y_2 - y_1)^2}$, and $CD = 4\sqrt{5}$,

$$\sqrt{(3 - a)^2 + (a + 1)^2} = 4\sqrt{5}$$
$$(3 - a)^2 + (a + 1)^2 = (4\sqrt{5})^2$$
$$2a^2 - 4a - 70 = 0$$
$$a^2 - 2a - 35 = 0$$
$$(a + 5)(a - 7) = 0$$
$$a = -5 \quad \text{or} \quad a = 7$$

Therefore, the two possible values of a are $a = -5$ or $a = 7$.

Example 3   Finding the coordinates of a vertex of a parallelogram

Three of the vertices of the parallelogram $ABCD$ are $A(5, 7)$, $B(-1, 5)$, and $C(-2, -3)$.

(a) Find the midpoint of AC.

(b) Find the coordinates of D.

Solution   Since $ABCD$ is a parallelogram, the midpoint of $\overline{AC}$ is the same as the midpoint of $\overline{BD}$.

$$\text{Midpoint of } \overline{AC} = \left(\frac{5 + -2}{2}, \frac{7 + -3}{2}\right) = \left(\frac{3}{2}, 2\right)$$

Let the coordinates of D be (x, y).

$$\text{Midpoint of } \overline{BD} = \left(\frac{-1 + x}{2}, \frac{5 + y}{2}\right)$$

Thus,

$$\text{Midpoint of } \overline{BD} = \text{Midpoint of } \overline{AC}$$
$$\left(\frac{-1 + x}{2}, \frac{5 + y}{2}\right) = \left(\frac{3}{2}, 2\right)$$

Solving $\dfrac{-1 + x}{2} = \dfrac{3}{2}$ and $\dfrac{5 + y}{2} = 2$ gives $x = 4$ and $y = -1$. Therefore, the coordinates of D of the parallelogram $ABCD$ is $(4, -1)$.

6.2 Finding areas of polygons using shoelace method

Finding areas of polygons using shoelace method

Suppose a triangle has vertices $A(x_1, y_1)$, $B(x_2, y_2)$, and $C(x_3, y_3)$, listed in clockwise order.

- Select one vertex: A

- Go clockwise and list all vertices ending with the same vertex: $A - B - C - A$

- Use the following formula called **shoelace** or **shoestring** method.

$$\frac{1}{2} \begin{vmatrix} x_1 & x_2 & x_3 & x_1 \\ y_1 & y_2 & y_3 & y_1 \end{vmatrix} = \frac{1}{2} \mid x_1 y_2 + x_2 y_3 + x_3 y_1 - y_1 x_2 - y_2 x_3 - y_3 x_1 \mid$$

For instance, for a triangle with vertices $A(1, 2)$, $B(7, 8)$ and $C(5, 5)$ listed in clockwise order,

$$\begin{aligned} \text{Area of triangle} &= \frac{1}{2} \begin{vmatrix} 1 & 7 & 5 & 1 \\ 2 & 8 & 5 & 2 \end{vmatrix} \\ &= \frac{1}{2} \mid 1(8) + 7(5) + 5(2) - 2(7) - 8(5) - 5(1) \mid \\ &= \frac{1}{2} \mid 8 + 35 + 10 - 14 - 40 - 5 \mid \\ &= \frac{1}{2} \mid -6 \mid \\ &= 3 \end{aligned}$$

Tip   Shoelace method can be extended for use with polygons with more than 3 sides.

Example 4   Finding the area of a pentagon using shoelace method

The vertices of the pentagon $ABCDE$ are $A(4,7)$, $B(7,-1)$, $C(5,-4)$, $D(-2,-2)$, and $E(-4,2)$ listed in clockwise order. Find the area of the pentagon $ABCDE$ using shoelace method.

Solution

$$
\begin{aligned}
\text{Area of pentagon} &= \frac{1}{2} \begin{vmatrix} 4 & 7 & 5 & -2 & -4 & 4 \\ 7 & -1 & -4 & -2 & 2 & 7 \end{vmatrix} \\
&= \frac{1}{2} \mid 4(-1) + 7(-4) + 5(-2) - 2(2) - 4(7) - 7(7) + 1(5) + 4(-2) + 2(-4) - 2(4) \mid \\
&= \frac{1}{2} \mid -4 - 28 - 10 - 4 - 28 - 49 + 5 - 8 - 8 - 8 \mid \\
&= \frac{1}{2} \mid -142 \mid \\
&= 71
\end{aligned}
$$

Therefore, the area of the pentagon $ABCDE$ is 71.

6.3 Linear law

Linear Law

Linear law is converting a non-linear equation to a linear equation given by

$$Y = mX + c$$

where:

- m is the gradient of graph of Y against X.

- c is the Y-intercept of the graph of Y against X.

- m and c are the constant.

- X and Y can be any expression involving x and y. Y and X can NOT contain coefficient and constant. For instance,

    (a) x^3 (3 is the exponent), $x\sqrt{x}$, xy, $\log y$, $x - y$, $\dfrac{x}{y}$, 2^x (2 is the base), are acceptable.

    (b) $4x^2$ (4 is the coefficient) and $xy - 5$ (5 is the constant) are NOT acceptable.

Example 5   Converting non-linear equations into the form $Y = mX + c$

Convert the following non-linear equations into the form $Y = mX + c$. State clearly what the variables X and Y and the constants m and c represent. (Note: there may be more than one way to convert it.)

(a) $y = a\sqrt{x} - \dfrac{b}{\sqrt{x}}$

(b) $y = ax^2 + bx$

(c) $y(x - a) = bx$

Solution

(a) $y = a\sqrt{x} - \dfrac{b}{\sqrt{x}}$. Multiplying each side of equation by $\sqrt{x}$ gives

$$y = a\sqrt{x} - \frac{b}{\sqrt{x}}$$
$$y\sqrt{x} = ax - b$$
$$Y = mX + c$$

Thus, $Y = y\sqrt{x}$, $X = x$, $m = a$, and $c = -b$.

(b) $y = ax^2 + bx$. Dividing each side of equation by x gives

$$y = ax^2 + bx$$
$$\frac{y}{x} = ax + b$$
$$Y = mX + c$$

Thus, $Y = \frac{y}{x}$, $X = x$, $m = a$, and $c = b$.

(c) $y(x - a) = bx$. Dividing each side of equation by y gives

$$y(x - a) = bx$$
$$x - a = b\frac{x}{y}$$
$$x = b\frac{x}{y} + a$$
$$Y = mX + c$$

Thus, $Y = x$, $X = \frac{x}{y}$, $m = b$, and $c = a$.

Example 6   Converting a nonlinear equation to a linear equation

Convert $y = ab^{\sqrt{x}}$, where a and b are constant, into the form $Y = mX + c$.

Solution   Taking common logarithms on both sides of the equation gives

$$y = ab^{\sqrt{x}}$$
$$\log_{10} y = \log_{10} ab^{\sqrt{x}}$$
$$\log_{10} y = \log_{10} a + \log_{10} b^{\sqrt{x}}$$
$$\log_{10} y = \log_{10} a + \sqrt{x} \log_{10} b$$
$$\log_{10} y = (\log_{10} b)\sqrt{x} + \log_{10} a$$
$$Y = mX + c$$

where $Y = \log_{10} y$, $m = \log_{10} b$, $X = \sqrt{x}$, and $c = \log_{10} a$.

Example 7   Finding relationship from a graph

For a relation shown below,

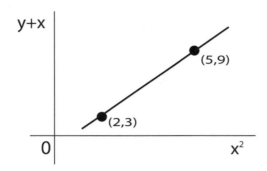

(a) Express y in terms of x.

(b) Find the value of y when $x = 3$.

Solution

(a) Let $Y = y + x$, and $X = x^2$. The gradient of graph of Y against X using two points $(2, 3)$ and $(5, 9)$ is $\dfrac{9 - 3}{5 - 2} = 2$. The Y-intercept of the graph of Y against X is -1. Thus, $Y = 2X - 1$. Replacing Y by $y + x$ and X by x^2 and solving for y gives

$$Y = 2X - 1$$
$$y + x = 2x^2 - 1$$
$$y = 2x^2 - x - 1$$

Therefore, y is terms of x is $y = 2x^2 - x - 1$.

(b) $y = f(x) = 2x^2 - x - 1$.

$$f(3) = 2(3)^2 - 3 - 1$$
$$= 14$$

Therefore, when $x = 3$, the value of y is 14.

EXERCISES

1. Find the equation of the line that is perpendicular to $y = -3x + 5$ and passes through the point $(-3, 6)$?

2. The line S has the equation $y = 2x - 5$. The line T has the equation $y = mx + b$. If two lines S and T are parallel and the distance between two points A and B is 4, find the values of m and b.

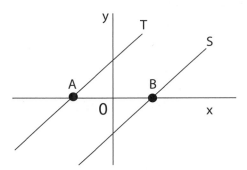

3. In the diagram below,

- The line AB has the equation $2x + y = 11$.

- The line AC has the equation $4x - y = 7$.

- The line BC has the equation $x - y = 4$.

- The lines AB and AC intersect at A.

- The lines AB and BC intersect at B.

- The lines AC and BC intersect at $C(1, -3)$.

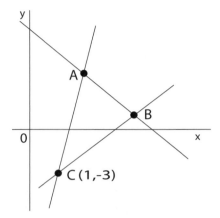

(a) Find the coordinates of points A and B.

(b) Calculate the length of the segment BC.

(c) Find the area of the triangle ABC bounded by lines AB, AC, and BC.

4. The table below shows values of the variables x and y.

x	1	2	3	4	5
y	-1	4	$\frac{23}{3}$	11	14.2

The variables are related by the equation $y = ax + \dfrac{b}{x}$.

(a) Copy and complete the following table.

x^2					
xy					

(b) Draw the graph of xy against x^2.

(c) Use your graph to find the values of a and b.

(d) Find the value of y when $x = 2.5$.

5. For each graph of the following relations, express y in terms of x.

   (a)

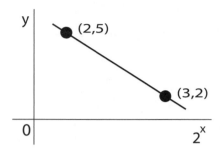

   (b)

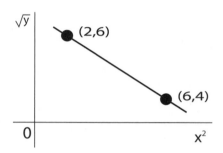

   (c)

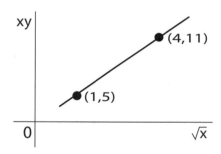

6. The curve $y = xy + x^2 - 1$ intersects the line $y = 2x + 1$ at the points A and B. Find the equation of the perpendicular bisector of the line AB.

7. Variables x and y are such that, when $\ln y$ is plotted against $\ln x$, a straight line graph passing through the points $(3, 3.4)$ and $(6, 1.4)$ is obtained.

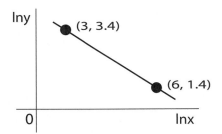

(a) Find the value of $\ln y$ when $\ln x = 0$.

(b) Given that $y = Ax^b$, find the values of A and b.

 **CHAPTER 7** Coordinate geometry of circles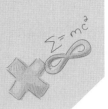

7.1 The standard equation of a circle

The standard equation of a circle

A **circle** is the set of all points that are equidistant from a fixed point called the center. The standard equation of a circle is given by

$$(x - h)^2 + (y - k)^2 = r^2$$

where the point (h, k) is the center of the circle, and r is the radius of the circle.

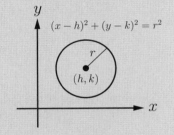

---

 1. In order to avoid a common mistake when finding the center of a circle, set $x - h = 0$ and $y - k = 0$ from the standard equation of a circle, and solve for x and y. Thus, the x- and y-coordinates of the center of the circle are $x = h$ and $y = k$.

2. If the equation of a circle is given by $x^2 + y^2 + 2gx + 2fx + c = 0$, use complete the square method to get the standard equation of a circle $(x - h)^2 + (y - k)^2 = r^2$.

Example 1   Finding the center and radius of a circle

Find the center and radius from the following equation of a circle.

(a) $x^2 + (y - 2)^2 = 7^2$

(b) $(x - 2)^2 + (y + 3)^2 = 25$

(c) $x^2 + y^2 + 6x - 8y - 11 = 0$

Solution

(a) $(x - 0)^2 + (y - 2)^2 = 7^2$. Set $x - 0 = 0$ and $y - 2 = 0$, and solve for x and y. Thus, the center of the circle is $(0, 2)$. The radius of the circle is 7.

(b) $(x - 2)^2 + (y + 3)^2 = 5^2$. Set $x - 2 = 0$ and $y + 3 = 0$, and solve for x and y. Thus, the center of the circle is $(2, -3)$. The radius of the circle is 5.

(c) Complete the square to get the standard equation of a circle $(x - h)^2 + (y - k)^2 = r^2$.

$$x^2 + y^2 + 6x - 8y - 11 = 0$$
$$(x^2 + 6x) + (y^2 - 8y) - 11 = 0$$
$$(x^2 + 6x + 0) + (y^2 - 8y + 0) - 11 = 0$$
$$(x^2 + 6x + 9 - 9) + (y^2 - 8y + 16 - 16) - 11 = 0$$
$$(x^2 + 6x + 9) + (y^2 - 8y + 16) - 9 - 16 - 11 = 0$$
$$(x + 3)^2 + (y - 4)^2 - 9 - 16 - 11 = 0$$
$$(x + 3)^2 + (y - 4)^2 = 6^2$$

Thus, the center of the circle is $(-3, 4)$. The radius of the circle is 6.

7.2 Intersection of a circle and a straight line

Solving Systems of Nonlinear Equations

A nonlinear equation represents a curve, not a straight line. Usually, the curve is either a circle, or a parabola. A system means more than one. Thus, a **system of nonlinear equations** contains at least one curve. Below is an example of a system of nonlinear equations.

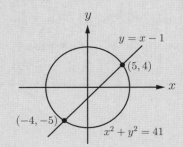

Solutions to a system of nonlinear equations are ordered pairs (x, y) that satisfy all equations in the system. In other words, solutions to a system of nonlinear equations are intersection points that lie on all graphs. In the figure above, $(5, 4)$ and $(-4, -5)$ are ordered pairs that satisfy all equations,

$$x^2 + y^2 = 41 \implies 5^2 + 4^2 = 41 \qquad x^2 + y^2 = 41 \implies (-4)^2 + (-5)^2 = 41$$
$$y = x - 1 \implies 4 = 5 - 1 \qquad y = x - 1 \implies -5 = -4 - 1$$

and are the **intersection points of a circle and a straight line**.

Solving a system of nonlinear equations means finding the x- and y-coordinates of the intersection points of both graphs. There are two methods to solve a system of nonlinear equations: substitution and elimination.

Let's solve the nonlinear equations shown below using the substitution method.

$$x^2 + y^2 = 41$$
$$y = x - 1$$

Substitute $x - 1$ for y into the first equation $x^2 + y^2 = 41$.

$x^2 + y^2 = 41$	Substitute $x - 1$ for y
$x^2 + (x - 1)^2 = 41$	Simplify
$2x^2 - 2x - 40 = 0$	Divide each side by 2
$x^2 - x - 20 = 0$	Factor
$(x + 4)(x - 5) = 0$	Solve for x
$x = -4$ or $x = 5$	

In order to find the values of y, substitute $x = 5$ and $x = -4$ into $y = x - 1$. Thus, $y = 4$ and $y = -5$, respectively. Therefore, the solutions to the system of nonlinear equations are $(5, 4)$ and $(-4, -5)$.

Intersection of a circle and a straight line

When a line $y = mx + b$ intersects a circle $(x - h)^2 + (y - k)^2 = r^2$, substitute $mx + b$ into y in $(x - h)^2 + (y - k)^2 = r^2$ and get the form $ax^2 + bx + c = 0$. The solutions to $ax^2 + bx + c = 0$ are the x-coordinates of the intersection points of the circle and the line.

Use the discriminant $\Delta = b^2 - 4ac$ to find the number of intersections points of the circle and the line.

- If $\Delta > 0$, there are two distinct intersection points.

- If $\Delta = 0$, there is only one intersection point (the line is tangent to the circle)

- If $\Delta < 0$, there are no intersection points.

Example 2   Finding the values of k when a line is tangent to a circle

Find the values of k for which the line $y = kx - 1$ is tangent to the circle $(x - 5)^2 + (y + 2)^2 = 8$.

Solution   Substituting $kx - 1$ into y in $(x - 5)^2 + (y + 2)^2 = 8$ gives

$$(x - 5)^2 + (y + 2)^2 = 8$$
$$(x - 5)^2 + (kx - 1 + 2)^2 = 8$$
$$(x - 5)^2 + (kx + 1)^2 = 8$$
$$x^2 - 10x + 25 + k^2 x^2 + 2kx + 1 = 8$$
$$(k^2 + 1)x^2 + (2k - 10)x + 18 = 0$$

Then use the discriminant to find the values of k. Since the line is tangent to the circle, the discriminant $\Delta = 0$.

$$b^2 - 4ac = 0$$
$$(2k - 10)^2 - 4(k^2 + 1)(18) = 0$$
$$-68k^2 - 40k + 28 = 0$$
$$17k^2 + 10k - 7 = 0$$
$$(k + 1)(17k - 7) = 0$$
$$k = -1 \quad \text{or} \quad k = \frac{7}{17}$$

Therefore, the values of k for which the line $y = kx - 1$ is tangent to the circle $(x - 5)^2 + (y + 2)^2 = 8$ are $k = -1$ or $k = \frac{7}{17}$.

7.3 The equation of a tangent line to a circle

Tangent line to a circle

A **tangent line** is a line that is drawn from outside of the circle and touches the circle at exactly one point. The point at which a tangent line touches the circle is the point of tangency.

A tangent line is **perpendicular** to the radius drawn to the point of tangency. In the figure to the right, A is the point of tangency and l is the tangent line to the circle. Thus, $\overline{OA} \perp l$.

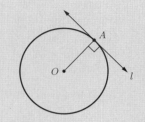

Example 3   **Finding the equation of a tangent line to a circle**

Find the equation of the line tangent to the circle $(x-2)^2 + (y-3)^2 = 13$ at the point $A(4,6)$ as shown below.

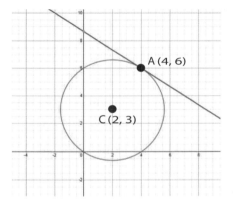

Solution   The slope of the radius drawn from the center $C(2,3)$ to the point of tangency $A(4,6)$ is $\dfrac{6-3}{4-2} = \dfrac{3}{2}$. The tangent line is perpendicular to the radius CA. So the slope of the tangent line is the negative reciprocal of $\dfrac{3}{2}$, which is $-\dfrac{2}{3}$. Writing the equation of the tangent line using point-slope form

$$y - 6 = -\frac{2}{3}(x - 4)$$
$$y = -\frac{2}{3}x + \frac{26}{3}$$

Therefore, the equation of the line tangent to the circle is $y = -\dfrac{2}{3}x + \dfrac{26}{3}$.

7.4 Intersection of two circles

Intersection of two circles

Two circles may intersect in two distinct points, a single point, or may not intersect. It can be determined by the following method.

Suppose the circle A has the radius of r_A, and the circle B has the radius of r_B. Calculate the distance, d, between center of the circle A and the center of the circle B.

- When $d > r_A + r_B$: two circles do not intersect.

- When $d = r_A + r_B$: two circles intersect at a single point.

- When $d < r_A + r_B$: two circles intersect at two distinct points.

Example 4   Finding the number of intersection points of two circles

The equations of two circles are given by $x^2 + y^2 + 2x + 4y - 4 = 0$ and $x^2 + y^2 - 6x - 4y + 9 = 0$. Find the number of intersection points of two circles.

Solution   Completing the square to the equation $x^2 + y^2 + 2x + 4y - 4 = 0$ gives

$$(x+1)^2 + (y+2)^2 = 3^2$$

with the center $(-1, -2)$ and the radius of 3. In addition, completing the square to the equation $x^2 + y^2 - 6x - 4y + 9 = 0$ gives

$$(x-3)^2 + (y-2)^2 = 2^2$$

with the center $(3, 2)$ and the radius of 2. The distance d between the center of the two circles is

$$d = \sqrt{(3 - (-1))^2 + (2 - (-2))^2} = 4\sqrt{2}$$

$d = 4\sqrt{2} \approx 5.67$. The sum of the radius of two circles is $3 + 2 = 5$. Since $d = 4\sqrt{2} > 5$, two circles do not intersect. Therefore, the number of intersection points of two circles is 0.

EXERCISES

1. For $k > 0$, the equation of the circle with the center $(k, -2k)$ and radius of 12 is given by

$$x^2 + y^2 - 2kx + 4ky - 4k^2 = 0$$

Find the value of k.

2. Find the points of intersection of the line $y = 4$ and the circle $x^2 + y^2 - 4x - 6y + 8 = 0$.

3. Find where the line $y = 2x + 6$ intersect the circle $(x + 5)^2 + (y - 1)^2 = 25$.

4. Find the equation of the line tangent to the circle $x^2 + y^2 + 2x - 4y - 36 = 0$ at the point $(-5, 7)$.

5. Find the points of intersection of the two circles given by

$$(x+4)^2 + y^2 = 49$$
$$x^2 + y^2 = 25$$

6. Find the range of the values of k for which the line $y = -2x + k$ intersects the circle $x^2 + y^2 - 6x + 4y - 32 = 0$ at two distinct points.

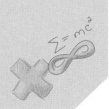

CHAPTER 8   Trigonometry

8.1 Circular measure

Angles

Two rays form an angle. One ray is called the **initial side** and the other ray is called the **terminal side**. An angle θ is in **standard position** if its vertex is at the origin and the initial side is on the positive x-axis shown in Figure 1.

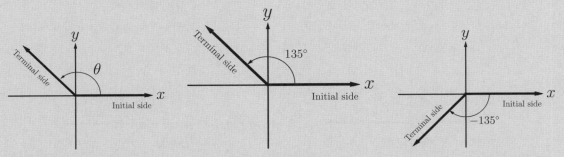

Fig. 1: Standard position   Fig. 2: Positive angle   Fig. 3: Negative angle

The angle shows the direction and amount of rotation from the initial side to the terminal side. If the rotation is in a counterclockwise direction, the angle is **positive** shown in Figure 2. Whereas, if the rotation is in a clockwise direction, the angle is **negative** shown in Figure 3.

Coterminal angle are angles in standard position that have a common terminal side. For instance, angles $135°$ and $-225°$, shown in Figure 4, are coterminal angles.

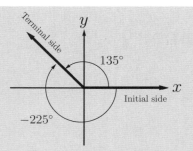

Fig. 4: Coterminal angles

Reference angle, β, is a positive acute angle formed by the terminal side and the closest x-axis, not the y-axis. If the terminal side lies in the first quadrant, reference angle β is the same as angle θ as shown in Figure 5. However, if the terminal side lies in other quadrants, reference angle β is different from the angle θ as shown in Figures 6, 7, and 8.

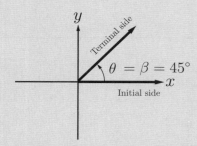

Fig. 5: When θ in the 1st quadrant

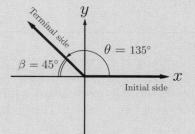

Fig. 6: When θ in the 2nd quadrant

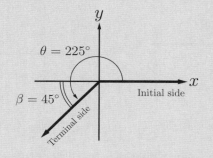

Fig. 7: When θ in the 3rd quadrant

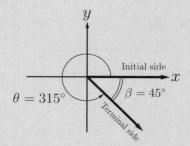

Fig. 8: When θ in the 4th quadrant

Arc length and Area of a sector

The radian is a very useful angular measure used in mathematics. Mathematicians prefer the radian to the degree (°) because it is a number that does not need an unit symbol. Although the radian can be denoted by the symbol "rad", it is usually omitted.

A **radian** is the measure of an angle θ at which the arc length is equal to the radius of the circle as shown in Figure 9.

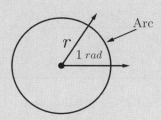

Fig. 9: At 1 radian, arc length = radius

Measuring one radian with a protractor, it is approximately 57.3°. Thus,

$$1\,rad \approx 57.3° \qquad \text{Multiply each side by } \pi = 3.141592\cdots$$
$$\pi\,rad = 180°$$

Since $\pi\,rad = 180°$, $2\pi\,rad = 360°$.

Arc length and area of a sector

An arc, shown in Figure 10, is a part of the circumference of a circle. A part can be expressed as the ratio of the central angle to 360° or 2π.

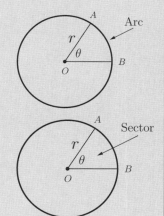

When θ (°) is given: $\quad$ Arc length $= 2\pi r \times \dfrac{\theta}{360°}$

When θ (rad) is given: $\quad$ Arc length $= 2\pi r \times \dfrac{\theta}{2\pi} = r\theta$ (rad)

A sector, shown in Figure 10, is a part of the area of a circle. A part can be expressed as the ratio of the central angle to 360° or 2π.

When θ (°) is given: $\quad$ Area of a sector $= \pi r^2 \times \dfrac{\theta}{360°}$

When θ (rad) is given: $\quad$ Area of a sector $= \pi r^2 \times \dfrac{\theta}{2\pi} = \dfrac{1}{2}r^2\theta$ (rad)

Figure 10

Conversion of Angles Between Degrees and Radians

The table below shows the conversion of angles between degrees and radians.

Degrees	30°	45°	60°	90°	120°	135°	150°	180°
Radians	$\frac{\pi}{6}$	$\frac{\pi}{4}$	$\frac{\pi}{3}$	$\frac{\pi}{2}$	$\frac{2\pi}{3}$	$\frac{3\pi}{4}$	$\frac{5\pi}{6}$	π

Degrees	210°	225°	240°	270°	300°	315°	330°	360°
Radians	$\frac{7\pi}{6}$	$\frac{5\pi}{4}$	$\frac{4\pi}{3}$	$\frac{3\pi}{2}$	$\frac{5\pi}{3}$	$\frac{7\pi}{4}$	$\frac{11\pi}{6}$	2π

1. To convert degrees to radians, multiply degrees by $\frac{\pi}{180°}$.

2. To convert radians to degrees, multiply radians by $\frac{180°}{\pi}$.

8.2 Finding the exact value of the trigonometric functions

Definitions of Six Trigonometric Functions

The six trigonometric functions are defined as ratios of sides in a right triangle. In the right triangle, shown to the right, the definitions of the six trigonometric functions are as follows:

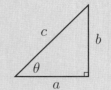

$$\sin\theta = \frac{\text{opposite}}{\text{hypotenuse}} = \frac{b}{c} \qquad \cos\theta = \frac{\text{adjacent}}{\text{hypotenuse}} = \frac{a}{c} \qquad \tan\theta = \frac{\text{opposite}}{\text{adjacent}} = \frac{b}{a}$$

$$\csc\theta = \frac{1}{\sin\theta} = \frac{c}{b} \qquad\qquad \sec\theta = \frac{1}{\cos\theta} = \frac{c}{a} \qquad\qquad \cot\theta = \frac{1}{\tan\theta} = \frac{a}{b}$$

 Note that $\tan\theta = \dfrac{\sin\theta}{\cos\theta}$ and $\cot\theta = \dfrac{\cos\theta}{\sin\theta}$.

Exact Value of Trigonometric Functions

The table below shows the exact value of $\sin\theta$, $\cos\theta$, and $\tan\theta$ at reference angle, β.

	Reference angle, β		
	$30° \left(\frac{\pi}{6}\right)$	$45° \left(\frac{\pi}{4}\right)$	$60° \left(\frac{\pi}{3}\right)$
$\sin\theta$	$\frac{1}{2}$	$\frac{\sqrt{2}}{2}$	$\frac{\sqrt{3}}{2}$
$\cos\theta$	$\frac{\sqrt{3}}{2}$	$\frac{\sqrt{2}}{2}$	$\frac{\sqrt{1}}{2}$
$\tan\theta$	$\frac{\sqrt{3}}{3}$	1	$\sqrt{3}$

Quadrant sign chart:

- II: $\sin\theta > 0$, $\csc\theta > 0$
- I: All positive
- III: $\tan\theta > 0$, $\cot\theta > 0$
- IV: $\cos\theta > 0$, $\sec\theta > 0$

The sign of a trigonometric function is determined by the quadrant the terminal side of the angle (or simply the angle) lies in. The chart above shows which quadrants the six trigonometric functions are positive. For instance, $\sin\theta$ and $\csc\theta$ are positive in the 1st and 2nd quadrants, $\cos\theta$ and $\sec\theta$ are positive in the 1st and 4th quadrants, and $\tan\theta$ and $\cot\theta$ are positive in the 1st and 3rd quadrants.

Evaluating trigonometric functions using the reference angle

When the angle θ lies in the either 2nd, or 3rd, or 4th quadrant, use the following formulas to evaluate the trigonometric functions.

$$\sin\theta = \pm\sin\beta \qquad\qquad \cos\theta = \pm\cos\beta \qquad\qquad \tan\theta = \pm\tan\beta$$

where β is the reference angle. Note that the sign of a trigonometric function, either $+$ or $-$, is determined by the quadrant the terminal side of the angle lies in.

Let's evaluate $\cos 225°$. As shown in the figure below, the angle $\theta = 225°$ lies in the third quadrant.

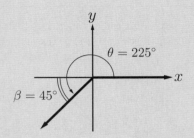

The reference angle β, an angle formed by the terminal side and the closest x-axis, is $45°$. Since the angle θ lies in the 3rd quadrant, the sign of the cosine function is negative. Thus,

$$\cos 225° = \pm \cos 45° \qquad \text{Since cosine is negative in the } 3^{\text{rd}} \text{ quadrant}$$

$$= -\cos 45° \qquad \text{Since } \cos 45° = \frac{\sqrt{2}}{2}$$

$$= -\frac{\sqrt{2}}{2}$$

Therefore, the exact value of $\cos 225°$ is $-\frac{\sqrt{2}}{2}$.

Example 1   Finding the quadrant where an angle θ lies

Find the quadrant where each of the following angles θ lies.

(a) If $\sin \theta < 0$ and $\cos \theta > 0$

(b) If $\tan \theta < 0$ and $\cos \theta < 0$

(c) If $\cot \theta > 0$ and $\csc \theta < 0$

Solution

(a) $\sin \theta < 0$ indicates that θ lies in the 3rd quadrant or 4th quadrant. Furthermore, $\cos \theta > 0$ indicates that θ lies in the 1st quadrant or 4th quadrant. Thus, θ must be in the 4th quadrant.

(b) $\tan \theta < 0$ indicates that θ lies in the 2nd quadrant or 4th quadrant. Furthermore, $\cos \theta < 0$ indicates that θ lies in the 2nd quadrant or 3rd quadrant. Thus, θ must be in the 2nd quadrant.

(c) $\cot \theta > 0$ indicates that θ lies in the 1st quadrant or 3rd quadrant. Furthermore, $\csc \theta < 0$ indicates that θ lies in the 3rd quadrant or 4th quadrant. Thus, θ must be in the 3rd quadrant.

Example 2   Finding exact values of trigonometric functions

If $\cos\theta = -\frac{3}{5}$ and $\sin\theta > 0$, find the exact value of each of the remaining trigonometric functions.

Solution $\cos\theta$ is negative in the 2nd and 3rd quadrants, and $\sin\theta$ is positive in the 1st and 2nd quadrants. Thus, θ must be in the 2nd quadrant. Since $\cos\theta = -\frac{3}{5}$, the length of the hypotenuse of a right triangle is 5 and the length of the adjacent side of θ is 3 as shown below. Thus, the length of the opposite side of θ is 4 using the Pythagorean theorem: $c^2 = a^2 + b^2$.

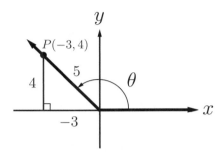

Suppose point $P(x, y)$ is on the terminal side of θ. Since θ lies in the 2nd quadrant, the x and y coordinates of point P is $(-3, 4)$. Use the definition of the trigonometric functions to find the exact values of the remaining trigonometric functions.

$$\sin\theta = \frac{\text{opposite}}{\text{hypotenuse}} = \frac{4}{5} \qquad \cos\theta = \frac{\text{adjacent}}{\text{hypotenuse}} = -\frac{3}{5} \qquad \tan\theta = \frac{\text{opposite}}{\text{adjacent}} = -\frac{4}{3}$$

$$\csc\theta = \frac{1}{\sin\theta} = \frac{5}{4} \qquad \sec\theta = \frac{1}{\cos\theta} = -\frac{5}{3} \qquad \cot\theta = \frac{1}{\tan\theta} = -\frac{3}{4}$$

Pythagorean Identities

$$\sin^2\theta + \cos^2\theta = 1 \qquad\qquad 1 + \tan^2\theta = \sec^2\theta \qquad\qquad 1 + \cot^2\theta = \csc^2\theta$$

 1. Note that $\sin^2\theta = (\sin\theta)^2$ and $\cos^2\theta = (\cos\theta)^2$.

2. The following variations of the Pythagorean identities are often used.

$$\sin^2\theta = 1 - \cos^2\theta, \qquad\qquad \cos^2\theta = 1 - \sin^2\theta$$

Even-Odd Properties

Knowing whether a trigonometric function is odd or even is useful when evaluating the trigonometric function of a negative angle. $\sin\theta$, $\csc\theta$, $\tan\theta$, and $\cot\theta$ are odd functions that satisfy $f(-\theta) = -f(\theta)$ for all θ. Whereas, $\cos\theta$ and $\sec\theta$ are even functions that satisfy $f(-\theta) = f(\theta)$ for all θ. Below is a summary of the even-odd properties for the six trigonometric functions.

$$\sin(-\theta) = -\sin\theta \qquad\qquad \csc(-\theta) = -\csc\theta$$
$$\cos(-\theta) = \cos\theta \qquad\qquad \sec(-\theta) = \sec\theta$$
$$\tan(-\theta) = -\tan\theta \qquad\qquad \cot(-\theta) = -\cot\theta$$

Example 3   Finding exact values using even-odd properties

Find the exact value of the following expressions.

(a) $\sin(-30°)$

(b) $\cos(-\frac{\pi}{6})$

Solution

(a) Since $\sin(-\theta) = -\sin\theta$, $\sin(-30°) = -\sin(30°) = -\frac{1}{2}$.

(b) Since $\cos(-\theta) = \cos\theta$, $\cos\left(-\frac{\pi}{6}\right) = \cos\left(\frac{\pi}{6}\right) = \frac{\sqrt{3}}{2}$.

Cofunctions Identities

Two angles are **complementary** if their sum is equal to 90°. The following cofunction identities show relationships between sine and cosine, tangent and cotangent, and secant and cosecant. The value of a trigonometric function of an angle is equal to the value of the cofunction of the complementary of the angle.

$$\sin(90° - \theta) = \cos\theta \qquad \tan(90° - \theta) = \cot\theta \qquad \sec(90° - \theta) = \csc\theta$$

For instance, 50° and 40° are complementary angles. Thus,

$$\sin 50° = \cos 40° \qquad \tan 50° = \cot 40° \qquad \sec 50° = \csc 40°$$

 Tip Since $180° = \pi$ radians, $90° = \frac{\pi}{2}$ radians. Thus, the cofunction identities can be expressed in radians.

$$\sin\left(\frac{\pi}{2} - \theta\right) = \cos\theta \qquad \tan\left(\frac{\pi}{2} - \theta\right) = \cot\theta \qquad \sec\left(\frac{\pi}{2} - \theta\right) = \csc\theta$$

8.3 Graphs of trigonometric functions

Graphs of Six Trigonometric Functions

Below shows the graphs of the six trigonometric functions: sine, cosine, tangent, cosecant, secant, and cotangent.

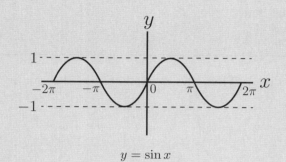

$$y = \sin x$$

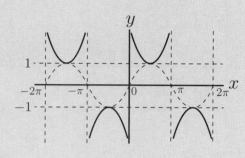

$$y = \csc x$$

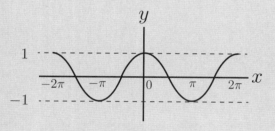

$$y = \cos x$$

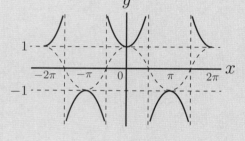

$$y = \sec x$$

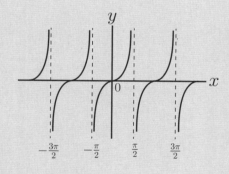

$$y = \tan x$$

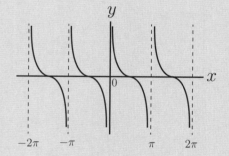

$$y = \cot x$$

Graphing the Sine and Cosine Functions

The general forms of the sine function and cosine function are as follows:

$$y = A \sin \big(B(x - C)\big) + D \qquad \text{or} \qquad y = A \cos \big(B(x - C)\big) + D$$

where A, B, C and D affect the amplitude, period, horizontal translation(horizontal shift), and vertical translation(vertical shift) of the graphs of the sine and cosine functions.

- A affects the amplitude. The amplitude is half the distance between the maximum and minimum values of the function. Since distance is always positive, the amplitude is $|A|$. For instance, both $y = 2 \sin x$ and $y = -2 \cos x$ have an amplitude of 2.

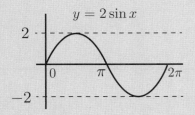
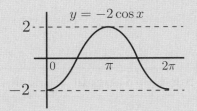

- B affects the period. The period, P, is the horizontal length of one complete cycle obtained by $P = \frac{2\pi}{B}$. For instance, the period of $y = \sin(2x)$ is $\frac{2\pi}{2} = \pi$, and the period of $y = \cos(\frac{1}{3}x)$ is $\frac{2\pi}{\frac{1}{3}} = 6\pi$.

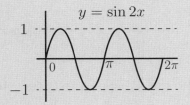
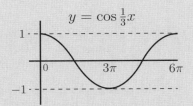

- C affects the horizontal translation. The horizontal translation is the measure of how far the graph has shifted horizontally. For instance, $y = \sin(x - \frac{\pi}{4})$ means a horizontal shift of $\frac{\pi}{4}$ to the right. Whereas, $y = \cos(x + \frac{\pi}{2})$ means a horizontal shift of $\frac{\pi}{2}$ to the left.

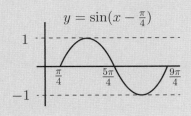
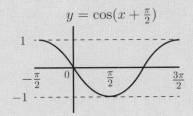

- D affects the vertical translation. The vertical translation is the measure of how far the graph has shifted vertically. For instance, $y = \sin x + 1$ means a vertical shift of 1 up. Whereas, $y = \cos x - 2$ means a vertical shift of 2 down.

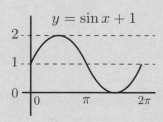

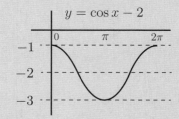

Graphing the Tangent Functions

The general form of the tangent functions is as follows:

$$y = A \tan \left(B(x - C) \right) + D$$

where B, C and D affect the period, horizontal translation(horizontal shift), and vertical translation(vertical shift) of the graphs of the tangent functions.

- The tangent function does not have an amplitude because it has no maximum or minimum value. If $|A| > 1$, the graph of the tangent function is vertically stretched. Whereas, if $0 < |A| < 1$, the graph of the tangent function is vertically compressed.

- B affects the period. The tangent function has the period of $\frac{\pi}{B}$ as opposed to the sine and cosine functions which have the period of $\frac{2\pi}{B}$. For instance, the period of $y = \tan(2x)$ is $\frac{\pi}{2}$. The graph of $y = \tan(2x)$ is shown in Figure 1.

- C affects the horizontal translation. For instance, $y = \tan(x - \frac{\pi}{2})$ means a horizontal translation of $\frac{\pi}{2}$ to the right. The graph of $y = \tan(x - \frac{\pi}{2})$ is shown in Figure 2.

- D affects the vertical translation. For instance, $y = \tan x + 3$ means a vertical translation of 3 up. The graph of $y = \tan x + 3$ is shown in Figure 3.

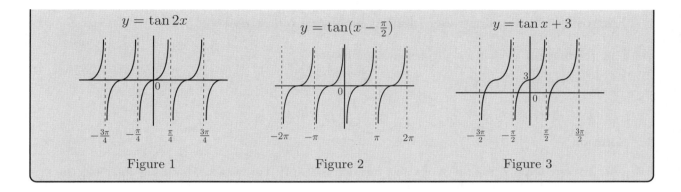

$y = \tan 2x$

$y = \tan(x - \frac{\pi}{2})$

$y = \tan x + 3$

Figure 1

Figure 2

Figure 3

Example 4   Finding amplitude, period, horizontal and vertical translation

Find the amplitude, period, horizontal and vertical translations of the following trigonometric functions.

(a) $y = 2\cos(x - \frac{\pi}{6}) + 1$

(b) $y = -3\sin(2x + \frac{\pi}{2}) - 2$

(c) $y = \tan(\frac{1}{2}x - \frac{\pi}{4}) + 3$

Solution

(a) Comparing $y = 2\cos(x - \frac{\pi}{6}) + 1$ to the general form of $y = A\cos\left(B(x - C)\right) + D$, we found that $A = 2$, $B = 1$, $C = \frac{\pi}{6}$, and $D = 1$. Thus, the amplitude is 2. The period, P, is $P = \frac{2\pi}{B} = \frac{2\pi}{1} = 2\pi$. The horizontal translation is $\frac{\pi}{6}$ to the right, and the vertical translation is 1 up.

(b) Change $y = -3\sin(2x + \frac{\pi}{2}) - 2$ to $y = -3\sin\left(2(x - (-\frac{\pi}{4}))\right) - 2$. Then, compare $y = -3\sin\left(2(x - (-\frac{\pi}{4}))\right) - 2$ to $y = A\sin\left(B(x - C)\right) + D$. We found that $A = -3$, $B = 2$, $C = -\frac{\pi}{4}$, and $D = -2$. Thus, the amplitude is $|-3| = 3$. The period, P, is $P = \frac{2\pi}{B} = \frac{2\pi}{2} = \pi$. The horizontal translation is $\frac{\pi}{4}$ to the left, and the vertical translation is 2 down.

(c) Change $y = \tan(\frac{1}{2}x - \frac{\pi}{4}) + 3$ to $y = \tan(\frac{1}{2}(x - \frac{\pi}{2})) + 3$. Then compare $y = \tan(\frac{1}{2}(x - \frac{\pi}{2})) + 3$ to $y = A\tan\left(B(x - C)\right) + D$. We found that $A = 1$, $B = \frac{1}{2}$, $C = \frac{\pi}{2}$, and $D = 3$. The tangent function does not have an amplitude because it has no maximum or minimum value. The period, P, is $P = \frac{\pi}{B} = \frac{\pi}{\frac{1}{2}} = 2\pi$. The horizontal translation is $\frac{\pi}{2}$ to the right, and the vertical translation is 3 up.

Example 5   Finding range of a trigonometric function

Find the range of the following trigonometric functions.

(a) $y = 2\sin(2x - 1) + 1$

(b) $y = -3\cos(x + \frac{\pi}{2}) - 2$

Solution

(a) For any angle x, the range of a sine function is $[-1, 1]$. For instance, both sine functions, $\sin x$ and $\sin(2x - 1)$ have the range of $[-1, 1]$.

$$-1 \le \sin(2x - 1) \le 1 \qquad\qquad \text{Multiply each side of inequality by 2}$$
$$-2 \le 2\sin(2x - 1) \le 2 \qquad\qquad \text{Add 1 to each side of inequality}$$
$$-1 \le 2\sin(2x - 1) + 1 \le 3$$

Thus, the range of $y = 2\sin(2x - 1) + 1$ is $[-1, 3]$.

(b) For any angle x, the range of a cosine function is $[-1, 1]$.

$$-1 \le \cos\left(x + \frac{\pi}{2}\right) \le 1 \qquad\qquad \text{Multiply each side of inequality by } -3$$

$$-3 \le -3\cos\left(x + \frac{\pi}{2}\right) \le 3 \qquad\qquad \text{Subtract 2 from each side of inequality}$$

$$-5 \le -3\cos\left(x + \frac{\pi}{2}\right) - 2 \le 1$$

Thus, the range of $y = -3\cos(x + \frac{\pi}{2}) - 2$ is $[-5, 1]$.

Example 6   Finding period from the graph of a trigonometric function

Graph $y = |\sin 2x|$, $0 \leq x \leq 2\pi$ and find the period of $|\sin 2x|$.

Solution   The period, P, of $y = \sin 2x$ is $P = \frac{2\pi}{2} = \pi$, which indicates that there are two complete cycles in the interval $[0, 2\pi]$ as shown in Figure 4.

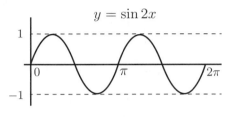

Figure 4

In order to graph $y = |\sin 2x|$, determine the part of the graph of $y = \sin 2x$ that lies below the x-axis as shown in Figure 5.

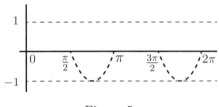

Figure 5

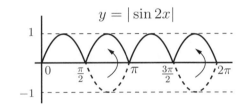

Figure 6

Lastly, reflect the part of the graph that lies below the x-axis about the x-axis as shown in Figure 6. Since the graph of $y = |\sin 2x|$ completes its cycle every $\frac{\pi}{2}$, the period $y = |\sin 2x|$ is $\frac{\pi}{2}$.

8.4 Solving trigonometric equations

Solving Trigonometric Equations

A trigonometric equation is an equation that involves trigonometric functions. The solutions of the trigonometric equation are the angles that satisfy the equation. In most cases, the Pythagorean Identities, factoring, the distributive property, or other algebraic skills are used to find the solutions of the equation.

Let's solve the trigonometric equation: $\sin \theta = -\frac{\sqrt{2}}{2}$, $0 \leq \theta < 2\pi$.

First, find the reference angle β for which $\sin \beta = \frac{\sqrt{2}}{2}$. The reference angle is $\beta = 45°$. Since the sine function is negative, the angle θ lies in either 3rd quadrant or 4th quadrant as shown below.

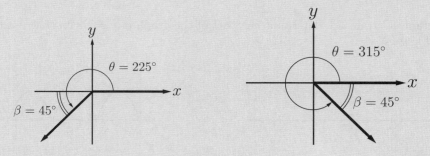

Therefore, the solutions of the equation are $225°$ (or $\frac{5\pi}{4}$) or $315°$ (or $\frac{7\pi}{4}$).

Example 7   Solving a trigonometric equation

Solve the equation: $2 \sin^2 \theta - 5 \sin \theta + 2 = 0$, $0 \leq \theta < 2\pi$.

Solution   Factor the expression $2 \sin^2 \theta - 5 \sin \theta + 2$ and solve for θ.

$$2 \sin^2 \theta - 5 \sin \theta + 2 = 0 \qquad \text{Factor}$$
$$(\sin \theta - 2)(2 \sin \theta - 1) = 0 \qquad \text{Solve}$$
$$\sin \theta = 2 \quad \text{or} \quad \sin \theta = \frac{1}{2}$$

Since $-1 \leq \sin \theta \leq 1$ for any angle θ, the equation $\sin \theta = 2$ has no solution. The solutions to $\sin \theta = \frac{1}{2}$, where $0 \leq \theta < 2\pi$ are $\frac{\pi}{6}$ and $\frac{5\pi}{6}$ or $30°$ and $150°$.

Example 8   Solving a trigonometric equation

Solve the equation: $\sin\theta + \cos\theta = 1, \ 0 \le \theta < 2\pi$.

Solution   Square each side of the equation.

$$\sin\theta + \cos\theta = 1$$
$$(\sin\theta + \cos\theta)^2 = 1$$
$$\sin^2\theta + 2\sin\theta\cos\theta + \cos^2\theta = 1 \qquad\qquad \sin^2\theta + \cos^2\theta = 1$$
$$2\sin\theta\cos\theta = 0$$

Thus, $\sin\theta = 0$ or $\cos\theta = 0$. Therefore, solutions to $\sin\theta + \cos\theta = 1$ are 0, π, $\frac{\pi}{2}$, and $\frac{3\pi}{2}$.

Example 9   Solving a trigonometric equation

Solve the equation: $2\cos^2\theta - 1 = \cos\theta, \ 0 \le \theta < 2\pi$.

Solution

$$2\cos^2\theta - 1 = \cos\theta$$
$$2\cos^2\theta - \cos\theta - 1 = 0$$
$$(2\cos\theta + 1)(\cos\theta - 1) = 0$$

Thus, $\cos\theta = -\frac{1}{2}$ or $\cos\theta = 1$. Therefore, solutions to $\cos 2\theta = \cos\theta$ are $\frac{2\pi}{3}$, $\frac{4\pi}{3}$ and 0.

8.5 Proving trigonometric identities

Proving Trigonometric Identities

In order to prove a trigonometric identity, you have to use logic steps to show that one side of the equation can be transformed to the other side of the equation. Do not work on both sides of the equation because you end up getting $1 = 1$. The following guidelines will help you prove trigonometric identities.

Guidelines

1. Start with one side of the equation that contains more complicated expressions(usually LHS).

2. Convert $\tan x$, $\cot x$, $\sec x$, and $\csc x$ to $\sin x$ or $\cos x$.

3. Use the Pythagorean identities.

$$\sin^2 x + \cos^2 x = 1, \qquad 1 - \cos^2 = \sin^2 x, \qquad 1 - \sin^2 = \cos^2 x$$
$$1 + \tan^2 x = \sec^2 x, \qquad \sec^2 x - 1 = \tan^2 x, \qquad \sec^2 x - \tan^2 x = 1$$
$$1 + \cot^2 x = \csc^2 x, \qquad \csc^2 x - 1 = \cot^2 x, \qquad \csc^2 x - \cot^2 x = 1$$

4. Expand the expression, combine like terms, and simplify the expression.

5. Factor the numerator and denominator using the factoring patterns shown below, then cancel common factors.

$$x^2 - y^2 = (x + y)(x - y)$$
$$x^2 \pm 2xy + y^2 = (x \pm y)^2$$
$$x^2 - 2x - 3 = (x + 1)(x - 3)$$
$$2x^2 - 5x + 2 = (x - 2)(2x - 1)$$

6. Show that LHS = RHS.

Example 10   Proving Trigonometric Identities

Prove the identity: $\dfrac{1 - \sin\theta}{\cos\theta} = \dfrac{\cos\theta}{1 + \sin\theta}$

Solution

$$
\begin{aligned}
\frac{1 - \sin\theta}{\cos\theta} &= \frac{1 - \sin\theta}{\cos\theta} \cdot \frac{1 + \sin\theta}{1 + \sin\theta} \\
&= \frac{1 - \sin^2\theta}{\cos\theta(1 + \sin\theta)} \\
&= \frac{\cos^2\theta}{\cos\theta(1 + \sin\theta)} \\
&= \frac{\cos\theta}{1 + \sin\theta}
\end{aligned}
$$

Example 11   Proving Trigonometric Identities

Prove the identity: $\dfrac{\sin\theta}{1 + \cos\theta} + \dfrac{1 + \cos\theta}{\sin\theta} = 2\csc\theta$

Solution

$$
\begin{aligned}
\frac{\sin\theta}{1 + \cos\theta} + \frac{1 + \cos\theta}{\sin\theta} &= \frac{\sin^2\theta + (1 + \cos\theta)^2}{\sin\theta(1 + \cos\theta)} \\
&= \frac{\sin^2\theta + 1 + 2\cos\theta + \cos^2\theta}{\sin\theta(1 + \cos\theta)} \\
&= \frac{\sin^2\theta + \cos^2\theta + 1 + 2\cos\theta}{\sin\theta(1 + \cos\theta)} \\
&= \frac{2 + 2\cos\theta}{\sin\theta(1 + \cos\theta)} \\
&= \frac{2(1 + \cos\theta)}{\sin\theta(1 + \cos\theta)} \\
&= \frac{2}{\sin\theta} \\
&= 2\csc\theta
\end{aligned}
$$

8.6 Area of non-right angled triangles

The Area of a SAS Triangles using $\sin\theta$

If triangle ABC shown at the right is a SAS triangle (a, b, and $m\angle C$ are known), the area of triangle ABC is as follows:

$$A = \frac{1}{2}ab\sin C$$

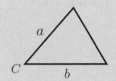

For instance, if triangle ABC shown at the right is a SAS triangle with $a = 5$, $b = 8$, and $m\angle C = 50°$, the area of triangle ABC is shown below.

$$A = \frac{1}{2}(5)(8)\sin 50° \approx 15.32$$

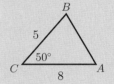

The Area of a SSS Triangles using Heron's formula

If triangle ABC shown at the right is a SSS triangle (a, b, c are known), the area of triangle ABC is as follows:

$$A = \sqrt{S(S-a)(S-b)(S-c)}$$

where S is the semi-perimeter and $S = \dfrac{a+b+c}{2}$.

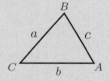

For instance, if triangle ABC shown at the right is a SSS triangle with $a = 2$, $b = 3$, and $c = 4$, $S = \dfrac{2+3+4}{2} = \dfrac{9}{2}$, and the area of triangle ABC is shown below.

$$A = \sqrt{\frac{9}{2}\left(\frac{9}{2}-2\right)\left(\frac{9}{2}-3\right)\left(\frac{9}{2}-4\right)} = \frac{3\sqrt{15}}{4}$$

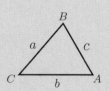

8.7 Solving triangles using the law of sines and cosines

Solving Triangles

Solving a triangle means finding the missing lengths of its sides and the measures of its angles. In general, solving a triangle can be classified as either solving a right triangle or solving a non-right triangle. The general rules for solving triangles are as follows:

- To solve a right triangle, use the definition of the six trigonometric functions or the Pythagorean theorem.

- To solve a non-right triangle, use the Law of Sines or Law of Cosines.

Solving Non-Right Triangles

Classifying non-right triangles

Depending on the information about the sides and angles given, non-right triangles can be classified as follows:

- ASA triangle: Two angles and the included side are known.

- SAA triangle: One side and two angles are known.

- SAS triangle: Two sides and the included angle are known.

- SSS triangle: Three sides are known.

The general rules for solving non-right triangles are as follows:

- To solve ASA and SAA triangles, use the Law of Sines.

- To solve SAS and SSS triangles, use the Law of Cosines.

 1. If none of the angles of a triangle is a right angle, the triangle is called either a non-right triangle or an oblique triangle.

2. SSA (Two sides and one angle opposite one of them) triangle is referred to as the **ambiguous case** because the given information may result in one triangle, two triangles, or no triangle.

The Law of Sines (ASA and SAA)

If a, b, and c are the lengths of the sides of a triangle, and A, B, and C are the opposite angles, then

$$\frac{a}{\sin A} = \frac{b}{\sin B} = \frac{c}{\sin C}$$

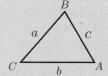

The Law of Sines implies that the largest angle is opposite the longest side and the smallest angle is opposite the shortest side. For instance, let a and c be the sides opposite the angles A and C of the triangle above, respectively. The Law of Sines satisfies the following:

$$\text{If } m\angle C < m\angle A \implies c < a$$
$$\text{If } c < a \implies m\angle C < m\angle A$$

Let's solve a non-right triangle ABC shown at the right. If $a = 8$, $m\angle A = 60°$ and $m\angle C = 40°$, find $m\angle B$, b, and c.

Since $m\angle A + m\angle B + m\angle C = 180°$, $m\angle B = 80$. Triangle ABC is a SAA triangle. Use the Law of Sines to find the sides b and c.

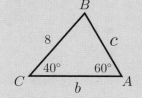

$$\frac{8}{\sin 60°} = \frac{b}{\sin 80°} \implies b = \frac{8\sin 80°}{\sin 60°} \approx 9.1$$

$$\frac{8}{\sin 60°} = \frac{c}{\sin 40°} \implies c = \frac{8\sin 40°}{\sin 60°} \approx 5.94$$

The Law of Cosines (SAS and SSS)

- If triangle ABC shown at the right is a SAS triangle (a, b, and $m\angle C$ are known), side c can be calculated by the Law of Cosines.

$$c^2 = a^2 + b^2 - 2ab\cos C$$

Note that side c is opposite angle C.

- If triangle ABC shown at the right is a SSS triangle (a, b, and c are known), the measure of angle A can be calculated by the Law of Cosines.

$$m\angle A = \cos^{-1}\left(\frac{a^2 - b^2 - c^2}{-2bc}\right)$$

Note that side a is opposite angle A.

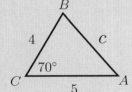

Let's find the side c and $m\angle B$ of a non-right triangle ABC shown at the right when $a = 4$, $b = 5$, and $m\angle C = 70°$.

Since triangle ABC is a SAS triangle, use the Law of Cosines to find the side c.

$$c^2 = 4^2 + 5^2 - 2(4)(5)\cos 70° \implies c \approx 5.23$$

To find the measure of angle B, use the Law of Cosines. Since side b is opposite angle B,

$$m\angle B = \cos^{-1}\left(\frac{b^2 - a^2 - c^2}{-2ac}\right) = \cos^{-1}\left(\frac{5^2 - 4^2 - 5.23^2}{-2(4)(5.23)}\right) \approx 63.98°$$

Example 12   Solving a ASA triangle using the Law of Sines

If $m\angle A = 55°$, $m\angle C = 45°$, and $b = 8$, solve the triangle shown below.

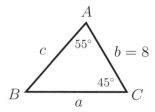

Solution   The sum of the measures of interior angles of a triangle is $180°$. Since $m\angle A = 55°$ and $m\angle C = 45°$, $m\angle B = 80°$. In order to find the sides a and c, use the Law of Sines.

$$\frac{8}{\sin 80°} = \frac{a}{\sin 55°} \implies a = \frac{8\sin 55°}{\sin 80°} \approx 6.65$$

$$\frac{8}{\sin 80°} = \frac{c}{\sin 45°} \implies c = \frac{8\sin 45°}{\sin 80°} \approx 5.74$$

Therefore, $m\angle B = 80°$, $a = 6.65$, and $c = 5.74$.

Example 13   Solving a SSS triangle using the Law of Cosines

Find the largest angle of a triangle under the given following conditions.

(a) if the ratio of the measures of three interior angles of the triangle is $2 : 3 : 4$.

(b) if the ratio of the three sides of the triangle is $2 : 3 : 4$.

Solution

(a) The ratio of the measures of three interior angles of the triangle is $2 : 3 : 4$. Let $2x$, $3x$, and $4x$ be the measures of the interior angles. Since the sum of the measures of interior angles of a triangle is $180°$,

$$2x + 3x + 4x = 180$$
$$9x = 180$$
$$x = 20$$

Therefore, the largest angle is $4x = 4(20) = 80°$.

(b) For simplicity, let the three sides a, b and c be 4, 3, and 2, respectively, since the ratio of the three sides is $2 : 3 : 4$. The Law of Sines implies that the largest angle is opposite the longest side. Thus, $\angle A$ is the largest angle since side a is the longest side.

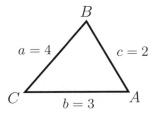

In order to find the measure of angle A, use the Law of Cosines. Since side a is opposite angle A,

$$m\angle A = \cos^{-1}\left(\frac{a^2 - b^2 - c^2}{-2bc}\right) = \cos^{-1}\left(\frac{4^2 - 3^2 - 2^2}{-2(3)(2)}\right) \approx 104.48°$$

Therefore, the largest angle of the triangle is $104.48°$.

EXERCISES

1. Convert the following angles to radians.

   (a) 120°

   (b) 225°

   (c) 270°

   (d) 330°

2. Convert the following radians to degrees.

   (a) $\dfrac{\pi}{5}$

   (b) $\dfrac{2\pi}{3}$

   (c) $\dfrac{5\pi}{12}$

   (d) $\dfrac{7\pi}{4}$

3. Given that $\cos\theta = \dfrac{\sqrt{2}}{3}$ and that θ is acute, find the exact values of the following expressions.

   (a) $\sin\theta$

   (b) $\tan\theta$

   (c) $\sin^2\theta$ (Note that $\sin^2\theta = (\sin\theta)^2$)

   (d) $\dfrac{3 + \sin\theta}{2 - \cos\theta}$

4. Given that $\tan\theta = \dfrac{3}{4}$ and that $\pi < \theta < \dfrac{3\pi}{2}$, find the exact value of the following expressions.

   (a) $\sin\theta$

   (b) $\cos\theta$

5. Given that $\cos A = \dfrac{2\sqrt{2}}{3}$ and $\tan B = -\dfrac{12}{5}$, where A and B are in the same quadrant, find the exact value of the following expressions.

   (a) $\sin A$

   (b) $\tan A$

   (c) $\sin B$

   (d) $\cos B$

6. Consider the function $f(x) = 3\cos 2x + 1$ for $0 \le x \le \pi$.

   (a) Find the amplitude.

   (b) Find the range of f.

   (c) Find the value of x does f reach its minimum.

   (d) Find the value of x does f reach its maximum.

   (e) Sketch the graph of f.

7. Sketch the graph of the following function for $0 \leq x \leq 360°$. State the range of each function.

    (a) $f(x) = |\tan x|$

    (b) $f(x) = |\cos 2x|$

    (c) $f(x) = |3\sin x - 2|$

8. The graph of a cosine function given by $y = a\cos bx + c$, where $a > 0$ has the given characteristics.

- Amplitude $= 2$
- Period $= \dfrac{2\pi}{3}$
- vertical shift $= -5$

Find the values of a, b and c.

9. Solve the following trigonometric equations on the interval $0 \leq \theta < 2\pi$.

    (a) $\cos \theta = -\dfrac{\sqrt{3}}{2}$

    (b) $\sin \theta = -\dfrac{\sqrt{2}}{2}$

    (c) $\cos \left(\theta - \dfrac{\pi}{6} \right) = -\dfrac{1}{2}$

    (d) $\sin 2\theta = \dfrac{\sqrt{3}}{2}$

10. Solve the following trigonometric equations on the interval $0 \leq \theta < 2\pi$.

    (a) $1 + \sin\theta = 2\cos^2\theta$

    (b) $\sin^2\theta = 2\cos\theta + 2$

    (c) $2\tan^2 x - 3\sec x + 3 = 0$

    (d) $2\cos^2\left(x - \dfrac{\pi}{6}\right) - \cos\left(x - \dfrac{\pi}{6}\right) = 1$

11. Prove the identity: $(\sin\theta + \cos\theta)^2 + (\sin\theta - \cos\theta)^2 = 2$.

12. Prove the identity: $1 - \dfrac{\cos^2\theta}{1 + \sin\theta} = \sin\theta$.

13. Prove the identity: $\dfrac{1 - \sin\theta}{\cos\theta} + \dfrac{\cos\theta}{1 - \sin\theta} = 2\sec\theta$.

14. Prove the identity: $\dfrac{1 - \sin\theta}{1 + \sin\theta} = (\sec\theta - \tan\theta)^2$.

15. In $\triangle ABC$ shown below, $AB = 7$, $BC = 5$, and $m\angle B = 75°$. Find AC.

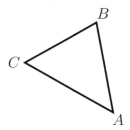

16. In $\triangle ABC$, $m\angle A : m\angle B : m\angle C = 2 : 3 : 4$, and $BC = 6$. Find AB.

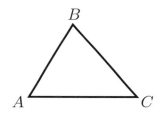

17. The circle has radius 9 cm and center O.

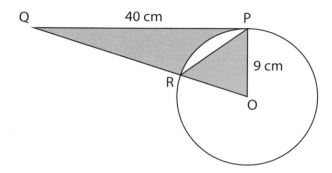

The segment QP is a tangent to the circle at the point P. $QP = 40$ cm and $PO = 9$ cm. The points Q, R, and O are collinear points.

(a) Find the area of the triangle QPO.

(b) Find the measure of the angle of QOP in radians. (Use a calculator)

(c) Find the length of PR.

(d) Find the area of the triangle OPR.

(e) Find the area of the shaded region.

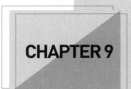 

CHAPTER 9 The binomial theorem

9.1 The Fundamental Counting Principle

Counting

Counting integers

How many positive integers are there between 42 and 97 inclusive? Are there 54, 55, or 56 integers? Even in this simple counting problem, many students are not sure what the right answer is. A rule for counting integers is as follows:

$$\text{The number of integers} = \text{Greatest integer} - \text{Least integer} + 1$$

According to this rule, the number of integers between 42 and 97 inclusive is $97 - 42 + 1 = 56$ integers.

Venn Diagram

A venn diagram is very useful in counting. It helps you count numbers correctly.

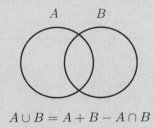

$$A \cup B = A + B - A \cap B$$

In the figure above, $A \cup B$ represents the combined area of two circles A and B. $A \cap B$ represents the common area where the two circles overlap. The venn diagram suggests that the combined area $(A \cup B)$ equals the sum of areas of circles $(A + B)$ minus the common area $(A \cap B)$.

In counting, each circle A and B represents a set of numbers. $n(A)$ and $n(B)$ represent the number of elements in set A and B, respectively. For instance, $A = \{2, 4, 6, 8, 10\}$ and $n(A) = 5$. Thus, the total number of elements that belong to either set A or set B, $n(A \cup B)$, can be counted as follows:

$$n(A \cup B) = n(A) + n(B) - n(A \cap B)$$

Let's find out how many positive integers less than or equal to 20 are divisible by 2 or 3. Define A as the set of numbers divisible by 2 and B as the set of numbers divisible by 3.

$$A = \{2, 4, 6, \cdots, 18, 20\}, \qquad n(A) = 10$$
$$B = \{3, 6, 9, 12, 15, 18\}, \qquad n(B) = 6$$
$$A \cap B = \{6, 12, 18\}, \qquad n(A \cap B) = 3$$

Notice that $A \cap B = \{6, 12, 18\}$ are multiples of 2 and multiples of 3. They are counted twice so they must be excluded in counting. Thus,

$$n(A \cup B) = n(A) + n(B) - n(A \cap B)$$
$$= 10 + 6 - 3$$
$$= 13$$

Therefore, the total number of positive integers less than or equal to 20 that are divisible by 2 or 3 is 13.

The Fundamental Counting Principle

- If event A can occur in m ways and event B can occur in n ways, then the number of ways both events A and B can occur is $m \times n$. For instance, Jason has three shirts and four pairs of jeans. He can dress up in $3 \times 4 = 12$ different ways.

- If there are n spaces to fill and m choices for each space, there are m^n ways. For instance, if there are 4 boxes and each box has 5 choices to choose a marble, there are $5 \times 5 \times 5 \times 5 = 5^4$ ways to put marbles into 4 boxes.

Addition Principle

Suppose two events A and B have no common outcomes. If event A can occur in m ways and another event can occur in n ways, and both events A and B cannot occur at the same time, then the number of ways that event A or B can occur is $m + n$. For instance, a bakery has a selection of 6 different cupcakes and 4 different donuts. If Jason selects only one treat, he has $6 + 4 = 10$ ways to choose from.

9.2 Permutation and combination

Permutation and Combination

Factorial notation

n factorial, denoted by $n!$, is defined as $n! = n(n-1)(n-2)\cdots 3\cdot 2\cdot 1$. In other words, n factorial is the product of all positive integers less than or equal to n. For instance, $3! = 3\cdot 2\cdot 1 = 6$. Below are the properties of factorials.

1. $0! = 1$ and $1! = 1$.

2. $n! = n\times(n-1)!$ or $n! = n\cdot(n-1)\cdot(n-2)!$.
   For instance, $5! = 5\cdot 4!$, or $5! = 5\cdot 4\cdot 3!$.

Permutations without repetition

A permutation, denoted by $_nP_r$, represents a number of ways to select r objects from the total number of objects n where the order is important. The permutation $_nP_r$ is given by

$$_nP_r = \frac{n!}{(n-r)!}, \qquad \text{where } r \leq n$$

For instance, how many words can be formed using all the letters in the word ABCDE?

Since all the letters A, B, C, D, and E are distinguishable, the order is important. Thus, this is a permutation problem.

$$_5P_5 = \frac{5!}{(5-5)!} = \frac{5!}{0!} = 120$$

Therefore, the number of different words can be formed using the letters in word ABCDE is 120.

Permutations with repetition

The number of permutations of n objects, where there are n_1 indistinguishable objects of one kind, and n_2 indistinguishable objects of a second kind, is given by

$$\text{Permutations with repetition} = \frac{n!}{n_1!\cdot n_2!}$$

For instance, how many words can be formed using all the letters in the word AABBB?

Since letters A and B are distinguishable, the order is important. However, there are 2 A's and 3 B's out of 5 letters.

$$\text{Permutations with repetition} = \frac{5!}{2!\cdot 3!} = \frac{5\cdot 4\cdot \cancel{3!}}{2!\cdot \cancel{3!}} = 10$$

Therefore, the number of different words can be formed using the letters in word AABBB is 10.

Combinations

A combination, denoted by $_nC_r$ or $\dbinom{n}{r}$, represents a number of ways to select r objects from the total number of objects n where the order is NOT important. The combination $\dbinom{n}{r}$ is given by

$$\binom{n}{r} = \frac{n!}{(n-r)! \cdot r!}, \qquad \text{where } r \leq n$$

For instance, how many 2 different books can be selected from a list of 10 books?

Since 2 books are indistinguishable, the order is not important. Thus, this is a combination problem.

$$\binom{10}{2} = \frac{10!}{8! \cdot 2!} = \frac{10 \cdot 9 \cdot \cancel{8!}}{\cancel{8!} \cdot 2!} = 45$$

Therefore, the number of selecting 2 different books from a list of 10 books is 45.

Example 1   Permutations and Combinations

(a) How many ways can a group of 10 people elect a president and a vice president?

(b) How many different committees of 2 officers can be formed from a group of 10 people?

Solution

(a) Since a president and a vice president are distinguishable, the order is important. Thus, this is a permutation problem.

$$_{10}P_2 = \frac{10!}{8!} = \frac{10 \cdot 9 \cdot \cancel{8!}}{\cancel{8!}} = 90$$

Therefore, the number of ways to elect a president and a vice present from a group of 10 people is 90.

(b) Since 2 officers are indistinguishable, the order is not important. Thus, this is a combination problem.

$$\binom{10}{2} = \frac{10!}{8! \cdot 2!} = \frac{10 \cdot 9 \cdot \cancel{8!}}{\cancel{8!} \cdot 2!} = 45$$

Therefore, the number of different committees of 2 officers can be formed from a group of 10 people is 45.

9.3 The binomial theorem

Binomial Expansions

$a + b$ is called a **binomial** as it contains two terms. Any expression of the term $(a + b)^n$ is called a **power of a binomial**. Let's consider the following algebraic expansions of the binomial $(a + b)^n$.

$$(a + b)^1 = a + b$$
$$(a + b)^2 = a^2 + 2ab + b^2$$
$$(a + b)^3 = (a + b)(a + b)^2 = a^3 + 3a^2b + 3ab^2 + b^3$$
$$(a + b)^4 = (a + b)(a + b)^3 = a^4 + 4a^3b + 6a^2b^2 + 4ab^3 + b^4$$

The algebraic expansions of $(a + b)^n$ get longer and longer as n increases. However, these algebraic expansions show some kind of pattern. The pattern is summed up by the **Binomial Theorem**.

Series

A **series** is the sum of terms (numbers or expressions). A series can be represented in a compact form, called summation notation or sigma notation $\sum$. Using the summation notation, the nth partial sum S_n can be expressed as follows:

$$S_n = a_1 + a_2 + \cdots + a_n = \sum_{k=1}^{n} a_k,$$

where k is called the **index** of the sum. $k = 1$ indicates where to start the sum and $k = n$ indicates where to end the sum. For instance,

$$\sum_{k=1}^{5} k^2 = 1^2 + 2^2 + 3^2 + 4^2 + 5^2$$

The Binomial Theorem

$$(a+b)^n = \sum_{k=0}^{n} \binom{n}{k} a^{n-k} b^k$$

$$(a-b)^n = (a+(-b))^n = \sum_{k=0}^{n} \binom{n}{k} a^{n-k} (-b)^k$$

where $\binom{n}{k} = \dfrac{n!}{(n-k)! \cdot k!}$ is the binomial coefficient which represents the number of combinations of n objects when k objects are taken at a time.

In general,

1. The number of terms of $(a+b)^n$ is $n+1$.

2. As the powers of a decreases by 1, the powers of b increases by 1.

3. The sum of the powers of a and b in each term of the expansion is n.

4. The general term, or $(r+1)$th term is $T_{r+1} = \binom{n}{r} a^{n-r} b^r$.

Example 2   Expand using the Binomial Theorem

Use the Binomial Theorem to expand $(x+2)^5$.

Solution

$$(x+2)^5 = \sum_{k=0}^{5} \binom{5}{k} x^{5-k} 2^k$$

$$= \binom{5}{0} x^5 (2^0) + \binom{5}{1} x^4 (2^1) + \binom{5}{2} x^3 (2^2) + \binom{5}{3} x^2 (2^3) + \binom{5}{4} x^1 (2^4) + \binom{5}{5} x^0 (2^5)$$

$$= x^5 + 5(2)x^4 + 10(4)x^3 + 10(8)x^2 + 5(16)x + 32$$

$$= x^5 + 10x^4 + 40x^3 + 80x^2 + 80x + 32$$

Example 3   Expand using the Binomial Theorem

Write down, in ascending powers of x, the first 4 terms in the expansion of $(2x + 1)^7$.

Solution   Rewrite $(2x + 1)^7$ as $(1 + 2x)^7$.

$$(1 + 2x)^7 = \sum_{k=0}^{7} \binom{7}{k} 1^{7-k}(2x)^k$$

$$= \binom{7}{0} 1^7 (2x)^0 + \binom{7}{1} 1^6 (2x)^1 + \binom{7}{2} 1^5 (2x)^2 + \binom{7}{3} 1^4 (2x)^3 + \cdots$$

$$= 1 + 14x + 84x^2 + 280x^3 + \cdots$$

Example 4   Finding the coefficient in a Binomial expansion

Find the coefficient of x^6 in the expansion of $(x^2 - 3)^4$.

Solution

$$(x^2 - 3)^4 = \sum_{k=0}^{4} \binom{4}{k} (x^2)^{4-k}(-3)^k$$

When $k = 1$, the term containing x^6 is $\binom{4}{1}(x^2)^3(-3)^1$ or $-12x^6$. Therefore, the coefficient of x^6 is -12.

EXERCISES

1. Evaluate the following expressions.

   (a) $_{10}P_2$

   (b) $_{8}P_5$

   (c) $_{7}C_3$

   (c) $\binom{9}{4}$

2. Find the value of n for which $\dfrac{n!}{(n-2)!} = 56$.

3. There are 5 light bulbs in a row. Each light bulb can be ON and OFF. How many different possibilities are there for the light bulbs?

4. There are 10 boys and 8 girls in Joshua's class. Joshua is going to select three different students in his class and write down their names in a list of order. The order will be boy-girl-boy or girl-boy-girl. How many different lists can Joshua write?

5. A committee of 4 is selected from 5 seniors and 6 juniors. Find the number of ways of selecting if it contains

   (a) 2 seniors and 2 juniors

   (b) At least 1 senior

6. There are three red chairs and two blue chairs in a room. Find the number of ways that you arrange these five chairs in a row.

7. There are ten points on a plane. If none of three points are collinear, find the followings.

   (a) How many segments connecting two points can be drawn?

   (b) How many triangles connecting three points can be drawn?

8. The letters, A, B, C, and D are to be arranged in a row. How many arrangements are possible if

   (a) they end with D?

   (b) begin with A and end with C?

9. Seven people enter a room and sit at random in a row of seven chairs. In how many ways can John, Daniel, and Joshua sit together in the row?

10. Expand and simplify.

   (a) $(x - y)^3$

   (b) $(2x + 3y)^3$

   (c) $\left(x - \dfrac{1}{x}\right)^4$

   (d) $(1 - 3x)^5$

11. Find the following terms.

    (a) the 5th terms of $(2x - 3)^7$

    (b) the 4th term of $(x^2 - 2x)^4$

    (c) the 7th term of $(x^2 + \frac{1}{x})^{10}$

12. Find the coefficient of following terms.

    (a) x^8 in the expansion of $(x^2 - 3)^6$

    (b) $a^3 b^4$ in the expansion of $(a + 2b^2)^5$

    (c) x^{10} in the expansion of $(2x - \frac{1}{x})^{10}$

13. The constant term in the expansion of $\left(x^4 - \dfrac{2}{x^3}\right)^{21}$ can be written as $a\dbinom{21}{b}$, where a is the integer. Find the values of a and b.

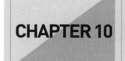

CHAPTER 10 Sequence and series

10.1 Sequence

Sequences

A sequence (or progression) is a list of numbers in order. The numbers in the list are called **terms** of the sequence and are denoted with subscripted letters: a_1 for the first term, a_2 for the second term, a_n for the nth term.

To evaluate the value of a_n, there are two types of formulas: the **explicit formula** and the **recursive formula**. The explicit formula evaluates a_n directly by substituting a value into the formula. Whereas, the recursive formula involves all previous terms to evaluate a_n. For instance, to evaluate the 10th term using the recursive formula, we need to evaluate the first nine terms.

Example 1   Evaluating the nth term using an explicit formula

If the sequence is defined by $a_n = 2n + 5$, evaluate the 5th term and the 11th term.

Solution   In order to evaluate the 5th term and the 11th term, substitute $n = 5$ and $n = 11$ into $a_n = 2n + 5$, respectively.

$$a_n = 2n + 5 \implies a_5 = 2(5) + 5 = 15$$
$$a_n = 2n + 5 \implies a_{11} = 2(11) + 5 = 27$$

Therefore, the value of a_5 is 15 and the value of a_{11} is 27.

167

Example 2 Evaluating the nth term using a recursive formula

If the sequence is defined by $a_n = 2a_{n-1} + 3$, $a_1 = 4$, evaluate the 5th term.

Solution In order to evaluate the 5th term, we need to find the previous four terms as shown below.

$$a_n = 2a_{n-1} + 3, \; a_1 = 4 \qquad \text{Recursive formula with } a_1 = 4$$
$$a_2 = 2a_1 + 3 = 2(4) + 3 = 11 \qquad \text{Substitute 2 for } n \text{ to find } a_2$$
$$a_3 = 2a_2 + 3 = 2(11) + 3 = 25 \qquad \text{Substitute 3 for } n \text{ to find } a_3$$
$$a_4 = 2a_3 + 3 = 2(25) + 3 = 53 \qquad \text{Substitute 4 for } n \text{ to find } a_4$$
$$a_5 = 2a_4 + 3 = 2(53) + 3 = 109 \qquad \text{Substitute 5 for } n \text{ to find } a_5$$

Therefore, the value of a_5 is 109.

Arithmetic Sequences and Geometric Sequences

There are two most common sequences: Arithmetic sequences and geometric sequences.

- In an **arithmetic sequence**, add or subtract the same number (common difference) to one term to get the next term.

- In a **geometric sequence**, multiply or divide one term by the same number (common ratio) to get the next term.

Type	Definition	Example	nth term
Arithmetic sequence	The common difference between any consecutive terms is constant.	$1, 3, 5, 7, \ldots$	$a_n = a_1 + (n-1)d$ where d is the common difference.
Geometric sequence	The common ratio between any consecutive terms is constant	$2, 4, 8, 16, \ldots$	$a_n = a_1 \times r^{n-1}$ where r is the common ratio.

Example 3   Writing the nth term of an arithmetic sequence

In the arithmetic sequence, if $a_9 = 51$ and $a_{17} = 99$, write an explicit formula for a_n.

Solution   Write the 9th term and 17th term of the arithmetic sequence in terms of a_1 and d using the nth term formula: $a_n = a_1 + (n-1)d$.

$$a_{17} = a_1 + 16d = 99$$
$$a_9 = a_1 + 8d = 51$$

Use the linear combinations method to solve for d and a_1.

$$a_1 + 16d = 99$$

$$\underline{a_1 + 8d = 51} \qquad \text{Subtract the two equations}$$

$$8d = 48 \qquad \text{Divide both sides by 8}$$

$$d = 6$$

Substitute $d = 6$ into $a_9 = a_1 + 8d = 51$ and solve for a_1. Thus, $a_1 = 3$. Therefore, the nth term of the arithmetic sequence is $a_n = a_1 + (n-1)d = 3 + (n-1)6 = 6n - 3$.

10.2 Series

Series

A **series** is the sum of a sequence. The finite series S_n is the sum of a finite number of terms. Whereas, the infinite series is the sum of an infinite number of terms. Often, the finite series S_n is called the **nth partial sum** and the infinite series is called an **infinite sum**.

The $(n-1)$th partial sum S_{n-1}, the nth partial sum S_n, and infinite sum S can be expressed as follows:

$$S_{n-1} = a_1 + a_2 + \cdots + a_{n-1} = \sum_{k=1}^{n-1} a_k,$$

$$S_n = a_1 + a_2 + \cdots + a_{n-1} + a_n = \sum_{k=1}^{n} a_k,$$

$$S = a_1 + a_2 + a_3 + \cdots = \sum_{k=1}^{\infty} a_k$$

Tip $a_n = S_n - S_{n-1}$

Arithmetic Series and Geometric Series

An arithmetic series is the sum of an arithmetic sequence. A geometric series is the sum of a geometric sequence. For instance, $1 + 3 + 5 + 7 + \cdots$ is an arithmetic series and $\frac{1}{2} + \frac{1}{4} + \frac{1}{8} + \frac{1}{16} + \cdots$ is a geometric series.

Below summarizes the nth partial sum and infinite sum for an arithmetic series and a geometric series.

Type	Arithmetic Series	Geometric Series
nth Partial Sum	$S_n = \frac{n}{2}(a_1 + a_n)$	$S_n = \frac{a_1(1-r^n)}{1-r}$
Infinite Sum	$S = \infty$	$S = \begin{cases} \dfrac{a_1}{1-r}, & \|r\| < 1 \\ \infty, & \|r\| \geq 1 \end{cases}$

Tip   1. $S_n = \dfrac{n}{2}(a_1 + a_n) = \dfrac{n}{2}(2a + (n-1)d)$

2. Note that the infinite sum S of a geometric series converges to $\dfrac{a_1}{1-r}$ if $|r| < 1$, where r is the common ratio of a geometric sequence.

Example 4   Finding the sum of an arithmetic series

Find the sum of the arithmetic series shown below.

$$3 + 7 + 11 + \cdots + 83$$

Solution $3, 7, 11, \cdots, 83$ is the arithmetic sequence with a common difference of 4. Using the nth term formula $a_n = a_1 + (n-1)d$,

$$
\begin{aligned}
a_n &= a_1 + (n-1)d && \text{Substitute 83 for } a_n, \text{ 3 for } a_1, \text{ and 4 for } d \\
83 &= 3 + (n-1)4 && \text{Subtract 3 from each side} \\
4(n-1) &= 80 && \text{Divide each side by 4} \\
n-1 &= 20 && \text{Add 1 to each side} \\
n &= 21
\end{aligned}
$$

we found that 83 is the 21st term of the arithmetic sequence. Thus, $3 + 7 + 11 + \cdots + 83$ is the sum of the first 21 terms of the arithmetic sequence.

$$
\begin{aligned}
S_n &= \frac{n}{2}(a_1 + a_n) && \text{Substitute 21 for } n \\
S_{21} &= \frac{21}{2}(a_1 + a_{21}) && \text{Substitute 3 for } a_1 \text{ and 83 for } a_{21} \\
&= \frac{21}{2}(3 + 83) = 903
\end{aligned}
$$

Therefore, $3 + 7 + 11 + \cdots + 83 = 903$.

Example 5   Finding the sum of a geometric series

Find the sum of each geometric series.

(a) $\frac{1}{2} + \frac{3}{4} + \frac{9}{8} + \cdots$

(b) $1 - \frac{1}{2} + \frac{1}{4} - \frac{1}{8} + \cdots$

(c) $\displaystyle\sum_{k=1}^{\infty} 4\left(-\frac{2}{3}\right)^{k-1}$

Solution

(a) The common ratio is

$$r = \frac{\frac{3}{4}}{\frac{1}{2}} = \frac{3}{2}$$

Since $|r| > 1$, the infinite sum S of the geometric series diverges. In other words, $S = \infty$.

(b) The common ratio is $-\frac{1}{2}$. Since $|r| < 1$,

$$S = \frac{a_1}{1-r} = \frac{1}{1-(-\frac{1}{2})} = \frac{1}{\frac{3}{2}} = \frac{2}{3}$$

Therefore, the infinite sum S of the geometric series is $\frac{2}{3}$.

(c) Substitute $k = 1$, $k = 2$, and $k = 3$ to write out the sum.

$$\sum_{k=1}^{\infty} 4\left(-\frac{2}{3}\right)^{k-1} = 4 - \frac{8}{3} + \frac{16}{9} + \cdots$$

The common ratio is $-\frac{2}{3}$. Since $|r| < 1$,

$$S = \frac{a_1}{1-r} = \frac{4}{1-(-\frac{2}{3})} = \frac{4}{\frac{5}{3}} = \frac{12}{5}$$

Therefore, the infinite sum S of the geometric series is $\frac{12}{5}$.

EXERCISES

1. Find k if $2k-1$, $k+5$, and $3k+2$ are three consecutive terms of an arithmetic sequence.

2. Find k if k, $2k-2$, and $2k+1$ are three consecutive geometric terms.

3. In the arithmetic sequence, the 5th term is 13 and 24th term is 89. Find the 16th term.

4. Find the sum of the series $\dfrac{2}{3} - 1 + \dfrac{3}{2} - \dfrac{9}{4} + \cdots$

5. The sum of the first n terms, S_n of an arithmetic progression is given by

$$S_n = 2n^2 + n$$

   (a) Find the expression for the nth term.

   (b) Find the value of a_{25}.

6. Answer the following questions.

   (a) Write the repeating decimal $63.6363\cdots$ as the geometric series.

   (b) Use your answer to part (a) to show that $63.6363\cdots$ can be expressed as $\dfrac{700}{11}$.

7. In the Fibonacci sequence, each number is the sum of two preceding terms. For instance, 1, 1, 2, 3, 5, 8, $\cdots$ forms the Fibonacci sequence. Suppose the first three terms of a different Fibonacci sequence are

$$a, \ 2b, \ a + 2b$$

Find the 7th term of the Fibonacci sequence in terms of a and b.

8. The sequence a_n is defined by $a_1 = 3$; $a_n = n^2 a_{n-1} - 1$. Find the value of a_3.

9. A ball is dropped from a height of 10 meters. Each time it hits the ground, it bounces to 80% of its previous height.

    (a) What height will the ball bounce up to after it strikes the ground for the third time?

    (b) Find the total vertical distance traveled by the tall.

10. A geometric progression has first term a, common ratio r and sum to n terms, S_n. Show that

$$\frac{S_{2n} - S_n}{S_n} = r^n$$

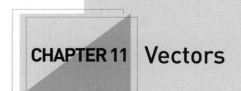

CHAPTER 11 Vectors

11.1 Vector notation

Vector notation

A vector is a quantity that has a magnitude and a direction. If point $A(x_1, y_1)$ is the initial point and point $B(x_2, y_2)$ is the terminal point, a vector, denoted by either $\overrightarrow{AB}$ or $\mathbf{AB}$, is defined as

$$\mathbf{AB} = \begin{bmatrix} x_2 - x_1 \\ y_2 - y_1 \end{bmatrix}$$

Suppose the vector $\mathbf{v} = \begin{bmatrix} a \\ b \end{bmatrix}$, where a is the horizontal component and b is the vertical components of

$\mathbf{v}$. $\begin{bmatrix} a \\ b \end{bmatrix}$ is called the **component form** of vector $\mathbf{v}$.

A **unit vector** is a vector of length 1 unit. The unit vector in the direction of positive x-axis is

$\mathbf{i} = \begin{bmatrix} 1 \\ 0 \end{bmatrix}$. The unit vector in the direction of positive y-axis is $\mathbf{j} = \begin{bmatrix} 0 \\ 1 \end{bmatrix}$. Another way to define vector

$\mathbf{v}$ in the form $a\mathbf{i} + b\mathbf{j}$ is as follows:

$$\mathbf{v} = \begin{bmatrix} a \\ b \end{bmatrix} = a \begin{bmatrix} 1 \\ 0 \end{bmatrix} + b \begin{bmatrix} 0 \\ 1 \end{bmatrix} = a\mathbf{i} + b\mathbf{j}$$

The magnitude (or modulus) of the vector of $\mathbf{v} = \begin{bmatrix} a \\ b \end{bmatrix}$, denoted by either $|\vec{v}|$ or $|\mathbf{v}|$, is defined as

$$|\mathbf{v}| = \sqrt{a^2 + b^2}$$

Unit Vector in the direction of the vector

The unit vector in the direction of $\mathbf{a}$ is defined as $\dfrac{\mathbf{a}}{|\mathbf{a}|}$.

- Vector $\mathbf{b}$ of length k, $k > 0$ in the same direction as $\mathbf{a}$ is $\mathbf{b} = k\dfrac{\mathbf{a}}{|\mathbf{a}|}$.

- Vector $\mathbf{b}$ of length k, $k > 0$ in the opposite direction as $\mathbf{a}$ is $\mathbf{b} = -k\dfrac{\mathbf{a}}{|\mathbf{a}|}$.

- Vector $\mathbf{b}$ of length k, $k > 0$ which is parallel to $\mathbf{a}$ is $\mathbf{b} = \pm k\dfrac{\mathbf{a}}{|\mathbf{a}|}$.

Example 1   Finding the unit vector

Find the unit vector in the direction of the following vectors.

(a) $3\mathbf{i} - 4\mathbf{j}$

(b) $-2\mathbf{i} - 3\mathbf{j}$

Solution

(a) Let $\mathbf{v} = 3\mathbf{i} - 4\mathbf{j}$. $|\mathbf{v}| = \sqrt{3^2 + (-4)^2} = 5$. Thus, the unit vector is

$$\frac{\mathbf{v}}{|\mathbf{v}|} = \frac{3\mathbf{i} - 4\mathbf{j}}{5} = \frac{3}{5}\mathbf{i} - \frac{4}{5}\mathbf{j} \qquad \text{or} \qquad \frac{\mathbf{v}}{|\mathbf{v}|} = \begin{bmatrix} \frac{3}{5} \\ -\frac{4}{5} \end{bmatrix}$$

(a) Let $\mathbf{v} = -2\mathbf{i} - 3\mathbf{j}$. $|\mathbf{v}| = \sqrt{(-2)^2 + (-3)^2} = \sqrt{13}$. Thus, the unit vector is

$$\frac{\mathbf{v}}{|\mathbf{v}|} = \frac{-2\mathbf{i} - 3\mathbf{j}}{\sqrt{13}} = -\frac{2}{\sqrt{13}}\mathbf{i} - \frac{3}{\sqrt{13}}\mathbf{j} \qquad \text{or} \qquad \frac{\mathbf{v}}{|\mathbf{v}|} = \begin{bmatrix} -\frac{2}{\sqrt{13}} \\ -\frac{3}{\sqrt{13}} \end{bmatrix}$$

11.2 Algebraic operations on vectors

Algebraic Operations on Vectors

If $\mathbf{u} = \begin{bmatrix} a \\ b \end{bmatrix}$, $\mathbf{v} = \begin{bmatrix} c \\ d \end{bmatrix}$, and c is a scalar (Constant), then

Addition: $\qquad \mathbf{u} + \mathbf{v} = \begin{bmatrix} a + c \\ b + d \end{bmatrix} = (a + c)\mathbf{i} + (b + d)\mathbf{j}$

Subtraction: $\qquad \mathbf{u} - \mathbf{v} = \begin{bmatrix} a - c \\ b - d \end{bmatrix} = (a - c)\mathbf{i} + (b - d)\mathbf{j}$

Scalar multiplication: $\qquad c\mathbf{u} = \begin{bmatrix} ca \\ cb \end{bmatrix} = ca\mathbf{i} + cb\mathbf{j}$

Example 2   Algebraic operations on vectors

If $\mathbf{v} = \begin{bmatrix} 3 \\ 4 \end{bmatrix}$, $\mathbf{w} = \begin{bmatrix} 2 \\ 8 \end{bmatrix}$, find $\mathbf{v} + \mathbf{w}$, $\mathbf{w} - \mathbf{v}$, $5\mathbf{v}$, and $|\mathbf{v} + \mathbf{w}|$.

Solution

$$\mathbf{v} + \mathbf{w} = \begin{bmatrix} 3 \\ 4 \end{bmatrix} + \begin{bmatrix} 2 \\ 8 \end{bmatrix} = \begin{bmatrix} 5 \\ 12 \end{bmatrix}$$

$$\mathbf{w} - \mathbf{v} = \begin{bmatrix} 2 \\ 8 \end{bmatrix} - \begin{bmatrix} 3 \\ 4 \end{bmatrix} = \begin{bmatrix} -1 \\ 4 \end{bmatrix}$$

$$5\mathbf{v} = 5 \begin{bmatrix} 3 \\ 4 \end{bmatrix} = \begin{bmatrix} 15 \\ 20 \end{bmatrix}$$

$$|\mathbf{v} + \mathbf{w}| = \sqrt{5^2 + 12^2} = 13$$

11.3 Vector geometry

Graphing Vectors

Vector equality

Two vectors **u** and **v** are equal if they have the same magnitude and the same direction. In Figure 1 below, three vectors **u**, **v**, and **w** are drawn. Although the three vectors have different initial points and different terminal points, they are equal because they have the same magnitude and the same direction.

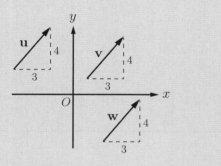

Figure 1

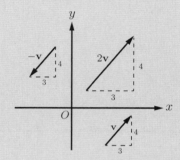

Figure 2

If **v** is a vector drawn in Figure 2, −**v** is a vector whose magnitude is the same as **v**, but whose direction is opposite to **v**. Whereas, 2**v** is a vector whose magnitude is twice the magnitude of **v**, but whose direction is the same as **v**.

Drawing the resultant vectors

The sum of two vectors is called the resultant vector. The resultant vector can be drawn in the following steps:

- Step 1: Draw the first vector **u** as shown in Figure 3.

- Step 2: Draw the second vector, **v**, so that the initial point of the second vector coincides with the terminal point of the first vector **u** as shown in Figure 3.

- Step 3: Draw the resultant vector, **u** + **v**, from the initial point of the first vector **u** to the terminal point of the second vector **v** as shown in Figure 4.

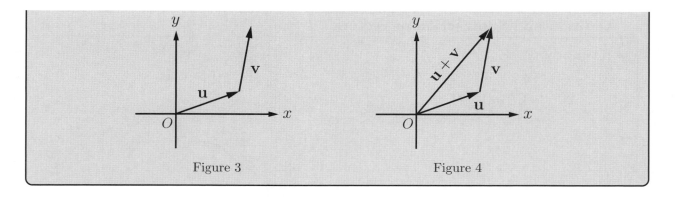

Figure 3   Figure 4

Parallelism

Two non-zero vectors are **parallel** if and only if one is a scalar multiple of the other.

- If **a** is parallel to **b**, then there exists a scalar k such that $\mathbf{a} = k\mathbf{b}$.

- If $\mathbf{a} = k\mathbf{b}$ for some scalar k, then **a** is parallel to **b** and $|\mathbf{a}| = |k||\mathbf{b}|$.

- If points A, B, and C are collinear points, $\overrightarrow{AB} = k\overrightarrow{AC}$ for some scalar k.

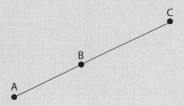

Addition and Subtraction of vectors

- Addition of vectors:

  (i) $\overrightarrow{PQ} + \overrightarrow{QR} = \overrightarrow{PR}$

  (ii) $\overrightarrow{PQ} + \overrightarrow{QR} + \overrightarrow{RS} = \overrightarrow{PS}$

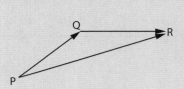
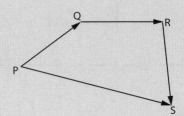

- Subtraction of vectors: $\overrightarrow{QR} = \overrightarrow{PR} - \overrightarrow{PQ}$

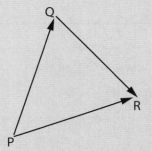

Expressing vectors in terms of two non-parallel vectors

If $\mathbf{a}$ and $\mathbf{b}$ are two non-zero and non-parallel vectors, any vector $\mathbf{c}$ in the plane containing $\mathbf{a}$ and $\mathbf{b}$ can be expressed as

$$\mathbf{c} = \lambda\mathbf{a} + \mu\mathbf{b}$$

is called a linear combination of vectors $\mathbf{a}$ and $\mathbf{b}$ with constants λ and μ.

- If $p\mathbf{a} + q\mathbf{b} = r\mathbf{a} + s\mathbf{b}$, then $p = r$ and $q = s$.

Example 3   Expressing vectors in terms of two non-parallel vectors

Let $\mathbf{a} = 3\mathbf{i} - 2\mathbf{j}$, $\mathbf{b} = 4\mathbf{i} + \mathbf{j}$, and $\mathbf{c} = -8\mathbf{i} - 13\mathbf{j}$. Find the λ and μ such that $\lambda\mathbf{a} + \mu\mathbf{b} = \mathbf{c}$.

Solution $\lambda\mathbf{a} + \mu\mathbf{b} = \mathbf{c}$.

$$\lambda(3\mathbf{i} - 2\mathbf{j}) + \mu(4\mathbf{i} + \mathbf{j}) = -8\mathbf{i} - 13\mathbf{j}$$
$$3\lambda\mathbf{i} - 2\lambda\mathbf{j} + 4\mu\mathbf{i} + \mu\mathbf{j} = -8\mathbf{i} - 13\mathbf{j}$$
$$(3\lambda + 4\mu)\mathbf{i} + (-2\lambda + \mu)\mathbf{j} = -8\mathbf{i} - 13\mathbf{j}$$

Equating the $\mathbf{i}$ gives $3\lambda + 4\mu = -8$, and equating the $\mathbf{j}$ gives $-2\lambda + \mu = -13$.

$$
\begin{array}{llll}
3\lambda + 4\mu = -8 & \xrightarrow{\phantom{\text{Multiply by } -4}} & 3\lambda + 4\mu = -8 \\
-2\lambda + \mu = -13 & \xrightarrow{\text{Multiply by } -4} & 8\lambda - 4\mu = 52
\end{array}
$$

Add two equations to eliminate μ variables.

$$
\begin{aligned}
3\lambda + 4\mu &= -8 \\
\underline{8\lambda - 4\mu} &= \underline{52} \qquad\qquad \text{Add two equations}\\
11\lambda &= 44 \\
\lambda &= 4
\end{aligned}
$$

Substituting $\lambda = 4$ into the first equation $3\lambda + 4\mu = -8$ gives $\mu = -5$. Therefore, $\lambda = 4$ and $\mu = -5$.

Example 4   Expressing vectors in terms of two non-parallel vectors

In the figure below, $OABC$ is a parallelogram.

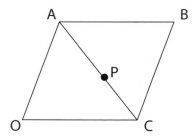

$\overrightarrow{OA} = \mathbf{a}$, and $\overrightarrow{OC} = \mathbf{b}$, The point P divides AC in the ratio 4:3. Write the expression for $\overrightarrow{OP}$ in terms of $\mathbf{a}$ and $\mathbf{b}$.

Solution

$$\overrightarrow{AC} = \overrightarrow{OC} - \overrightarrow{OA} = \mathbf{b} - \mathbf{a}$$

Since the point P divides AC in the ratio 4:3, $\overrightarrow{AP} = \frac{4}{7}\overrightarrow{AC}$.

$$\overrightarrow{AP} = \frac{4}{7}\overrightarrow{AC} = \frac{4}{7}(\mathbf{b} - \mathbf{a})$$

Since $\overrightarrow{OP} = \overrightarrow{OA} + \overrightarrow{AP}$,

$$\begin{aligned}\overrightarrow{OP} &= \overrightarrow{OA} + \overrightarrow{AP} \\ &= \mathbf{a} + \frac{4}{7}(\mathbf{b} - \mathbf{a}) \\ &= \frac{3}{7}\mathbf{a} + \frac{4}{7}\mathbf{b} \\ &= \frac{1}{7}(3\mathbf{a} + 4\mathbf{b})\end{aligned}$$

Therefore, $\overrightarrow{OP}$ in terms of $\mathbf{a}$ and $\mathbf{b}$ is $\frac{1}{7}(3\mathbf{a} + 4\mathbf{b})$.

Example 5 — Expressing vectors in terms of two non-parallel vectors

In the figure below, $\overrightarrow{OA} = \mathbf{a}$ and $\overrightarrow{OB} = \mathbf{b}$.

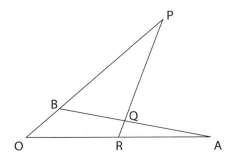

R is the midpoint of OA and $\overrightarrow{OP} = 4\overrightarrow{OB}$. $\overrightarrow{AQ} = \lambda\overrightarrow{AB}$ and $\overrightarrow{RQ} = \mu\overrightarrow{RP}$.

(a) Find $\overrightarrow{OQ}$ in terms of λ, $\mathbf{a}$, and $\mathbf{b}$.

(b) Find $\overrightarrow{OQ}$ in terms of μ, $\mathbf{a}$, and $\mathbf{b}$.

(c) Find the values of λ and μ.

(d) Find the ratio of BQ to QA.

Solution

(a) $\overrightarrow{OR} = \dfrac{1}{2}\overrightarrow{OA} = \dfrac{1}{2}\mathbf{a}$.

$$\overrightarrow{AB} = \overrightarrow{OB} - \overrightarrow{OA} = \mathbf{b} - \mathbf{a}$$
$$\overrightarrow{AQ} = \lambda\overrightarrow{AB} = \lambda(\mathbf{b} - \mathbf{a}) = \lambda\mathbf{b} - \lambda\mathbf{a}$$
$$\overrightarrow{OQ} = \overrightarrow{OA} + \overrightarrow{AQ} = \mathbf{a} + \lambda\mathbf{b} - \lambda\mathbf{a} = (1-\lambda)\mathbf{a} + \lambda\mathbf{b}$$

(b) $\overrightarrow{OP} = 4\overrightarrow{OB} = 4\mathbf{b}$.

$$\overrightarrow{RP} = \overrightarrow{OP} - \overrightarrow{OR} = 4\mathbf{b} - \dfrac{1}{2}\mathbf{a}$$
$$\overrightarrow{RQ} = \mu\overrightarrow{RP} = \mu(4\mathbf{b} - \dfrac{1}{2}\mathbf{a}) = -\dfrac{1}{2}\mu\mathbf{a} + 4\mu\mathbf{b}$$
$$\overrightarrow{OQ} = \overrightarrow{OR} + \overrightarrow{RQ} = \dfrac{1}{2}\mathbf{a} - \dfrac{1}{2}\mu\mathbf{a} + 4\mu\mathbf{b} = \left(\dfrac{1}{2} - \dfrac{1}{2}\mu\right)\mathbf{a} + 4\mu\mathbf{b}$$

(c) From the part (a) and (b), $\overrightarrow{OQ} = (1-\lambda)\mathbf{a} + \lambda\mathbf{b} = \left(\dfrac{1}{2} - \dfrac{1}{2}\mu\right)\mathbf{a} + 4\mu\mathbf{b}$. Equating $\lambda\mathbf{b} = 4\mu\mathbf{b}$ gives

$\lambda = 4\mu$. In addition, equating $(1 - \lambda)\mathbf{a} = \left(\frac{1}{2} - \frac{1}{2}\mu\right)\mathbf{a}$ and $\lambda = 4\mu$ gives,

$$1 - \lambda = \frac{1}{2} - \frac{1}{2}\mu$$
$$2 - 2\lambda = 1 - \mu$$
$$2 - 2(4\mu) = 1 - \mu$$
$$2 - 8\mu = 1 - \mu$$
$$-7\mu = -1$$
$$\mu = \frac{1}{7}$$

Since $\lambda = 4\mu$, $\lambda = \frac{4}{7}$. Therefore, the values of $\lambda = \frac{4}{7}$ and the value of $\mu = \frac{1}{7}$.

(d) Since $\overrightarrow{AQ} = \frac{4}{7}\overrightarrow{AB}$, the ratio of BQ to QA is $3 : 4$.

11.4 Constant velocity problems

Relative Position of a Vector

In the figure below, the position of a point A with respect to the origin O is indicated by the directed line segment $\overrightarrow{OA}$, or $\mathbf{r_A}$. The vectors $\overrightarrow{OA}$, or $\mathbf{r_A}$ are called **position vector** of A relative to the origin O.

If A has a position vector $\mathbf{r_A}$ and B has position vector $\mathbf{r_B}$ as shown below,

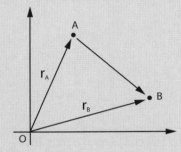

then the **relative position** of B from A can be expressed as $\overrightarrow{AB} = \overrightarrow{OB} - \overrightarrow{OA}$, or $\overrightarrow{AB} = \boldsymbol{r_B} - \boldsymbol{r_A}$.

Application of Vectors

Physics is the study in which vectors are used the most. The following quantities are distinguished as either scalar or vector quantities.

- **Displacement** is a vector quantity since it has a direction and magnitude.

- **Distance** is a scalar and it is the magnitude of displacement.

- **Velocity** is a vector quantity since it has a direction and magnitude.

- **Speed** is a scalar and it is the magnitude of velocity.

Example 6   Finding a vector velocity

A particle travels at a constant velocity from point A to B with $\overrightarrow{AB} = 25\mathbf{i} - 60\mathbf{j}$ m, and the time taken is 5s. Find the velocity vector and the speed of the particle.

Solution

$$\text{Velocity} = \frac{\text{Displacement}}{\text{Time taken}} = \frac{25\mathbf{i} - 60\mathbf{j}}{5} = 5\mathbf{i} - 12\mathbf{j} \ \text{ms}^{-1}$$
$$\text{Speed} = \sqrt{5^2 + (-12)^2} = 13 \ \text{ms}^{-1}$$

Splitting a velocity vector into its components

The velocity $\mathbf{v}$ of a particle traveling with magnitude r can be written in the form $x\mathbf{i} + y\mathbf{j}$.

$$x = r\cos\theta, \qquad y = r\sin\theta$$

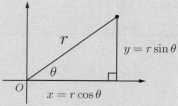

where θ is the positive angle formed by the positive x-axis and the terminal side of $\mathbf{v}$. Thus, $\mathbf{v} = r\cos\theta\mathbf{i} + r\sin\theta\mathbf{j}$.

For instance, the velocity $\mathbf{v}$ of a particle traveling on a bearing of $315°$ at 10 ms^{-1} as shown in Figure 5. The positive angle θ formed by the positive x-axis and the terminal side of $\mathbf{v}$ is $135°$ as shown in Figure 6.

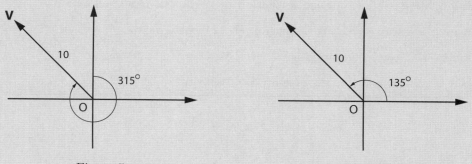

Figure 5   Figure 6

Thus, the horizontal component of $\mathbf{v}$ is $x = 10\cos 135° = -5\sqrt{2}$, and the vertical component of $\mathbf{v}$ is $y = 10\sin 135° = 5\sqrt{2}$. Therefore, $\mathbf{v} = x\mathbf{i} + y\mathbf{j} = -5\sqrt{2}\mathbf{i} + 5\sqrt{2}\mathbf{j}$.

The position vector of an object at time t

If an object has initial position vector $\mathbf{a}$ and moves with a constant velocity $\mathbf{v}$, the position vector $\mathbf{r}$ at time t is given by

$$\mathbf{r} = \mathbf{a} + t\mathbf{v}$$

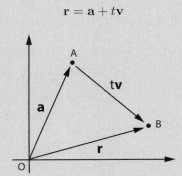

Example 7   Finding the position vector

A car travels from a point P with position vector $(50\mathbf{i} - 40\mathbf{j})$ km to a point Q. The car travels with a constant velocity $(-5\mathbf{i} + 10\mathbf{j})$ km h^{-1} and takes 3 hours to complete the journey. Find the position vector of the point Q.

Solution   Suppose the position vector of the point Q is $\mathbf{r}$.

$$\begin{aligned}
\mathbf{r} &= \mathbf{a} + t\mathbf{v} \\
&= 50\mathbf{i} - 40\mathbf{j} + 3(-5\mathbf{i} + 10\mathbf{j}) \\
&= 50\mathbf{i} - 40\mathbf{j} - 15\mathbf{i} + 30\mathbf{j} \\
&= 35\mathbf{i} - 10\mathbf{j}
\end{aligned}$$

Therefore, the position vector of the point Q is $(35\mathbf{i} - 10\mathbf{j})$ km.

EXERCISES

1. Find the following vectors.

   (a) Find the unit vector in the direction of $-20\mathbf{i} + 21\mathbf{j}$.

   (b) Find the vector $\mathbf{v}$ in the direction of $2\mathbf{i} - 3\mathbf{j}$ with magnitude 25.

2. Find the following velocity vector in the form $x\mathbf{i} + y\mathbf{j}$.

   (a) The velocity vector $\mathbf{v}$ of a particle traveling on a bearing of $120°$ at 100 ms^{-1}.

   (b) The velocity vector $\mathbf{v}$ of a particle traveling on a bearing of $210°$ at 42 ms^{-1}.

3. Relative to an origin O, the position vector A is $-11\mathbf{i} + 7\mathbf{j}$ and the position vector of B is $4\mathbf{i} + 2\mathbf{j}$.

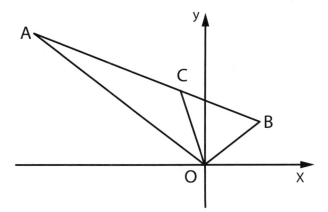

(a) Find $\overrightarrow{BA}$.

(b) The point C lies on AB such that $\overrightarrow{BC} = \dfrac{2}{5}\overrightarrow{BA}$. Find the position vector of C.

4. In the diagram below, $\overrightarrow{AD} = 3\mathbf{a}$, $\overrightarrow{BC} = 2\mathbf{a}$, and $\overrightarrow{CD} = \mathbf{b}$.

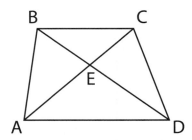

$\overrightarrow{AE} = \lambda\overrightarrow{AC}$ and $\overrightarrow{BE} = \mu\overrightarrow{BD}$.

(a) Find $\overrightarrow{AB}$ in terms of $\mathbf{a}$ and $\mathbf{b}$.

(b) Find $\overrightarrow{BD}$ in terms of $\mathbf{a}$ and $\mathbf{b}$.

(c) Find $\overrightarrow{AE}$ in terms of λ, μ, $\mathbf{a}$ and/or $\mathbf{b}$.

(d) Find $\overrightarrow{BE}$ in terms of λ, μ, $\mathbf{a}$ and/or $\mathbf{b}$.

(e) Find the value of λ and the value of μ.

5. Let $\mathbf{a} = 5\mathbf{i} - 6\mathbf{j}$, $\mathbf{b} = -1\mathbf{i} + 2\mathbf{j}$, and $\mathbf{c} = -13\mathbf{i} + 18\mathbf{j}$. Find the λ and μ such that $\lambda\mathbf{a} + \mu\mathbf{b} = \mathbf{c}$.

6. At 1 PM, a car travels from the point A and travels a distance of 90 km to the point B. The position vector, $\mathbf{r}$ km, of the car relative to an origin O, t hours after 1 PM is given by

$$\mathbf{r} = \begin{bmatrix} -10 + 6t \\ 40 + 8t \end{bmatrix}$$

(a) Find the position vector of the point A.

(b) Find the velocity vector of the car.

(c) Find the position vector of the point B.

7. Two boats, A and B, are spotted from a lighthouse. The courses of the two boats have the following equations.

$$\mathbf{A} = (-1 + 2t)\mathbf{i} + (10 - 2t)\mathbf{j}$$
$$\mathbf{B} = 2s\mathbf{i} + (4 + 3s)\mathbf{j}$$

All units are given in km and the time in hours.

(a) Find the positions of the boats when they were spotted.

(b) Find the distance between two boats when they were spotted.

(c) Find the speed of each boat.

(d) Find the coordinates of the point where the courses meet.

(e) Determine whether two boats will collide or not.

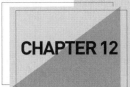

CHAPTER 12 Derivative functions

12.1 Instantaneous rate of change

Rate of Change

Average rate change measures how much f changes over an interval from $x = a$ to $x = a + h$. Thus, average rate of change is defined as

$$\text{Average rate of change} = \frac{\Delta f}{\Delta x} = \frac{f(a+h) - f(a)}{h}$$

Notice that the average rate of change is the slope of the secant line in Figure 1. Whereas, **instantaneous rate of change** measures how much f changes over a very short interval$(h \approx 0)$. Thus, instantaneous rate of change is defined as

$$\text{Instantaneous rate of change} = \lim_{h \to 0} \frac{\Delta f}{\Delta x} = \lim_{h \to 0} \frac{f(a+h) - f(a)}{h}$$

Notice that the instantaneous rate of change is the slope of the tangent line in Figure 2.

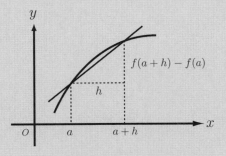

Figure 1

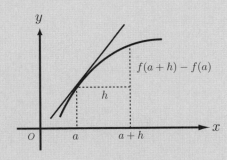

Figure 2

1. A **secant line** is a straight line that joins two points on a function.

2. A **tangent line** is a straight line that touches the function at a point without crossing over.

3. Instantaneous rate of change can be expressed by one of the two forms shown below.

$$\lim_{h \to 0} \frac{f(a+h) - f(a)}{h} \qquad \text{or} \qquad \lim_{x \to a} \frac{f(x) - f(a)}{x - a}$$

Definition of the Derivative Function

The **derivative function of f**, **gradient function of f**, or **slope function of f** is defined as

$$f'(x) = \lim_{h \to 0} \frac{f(x+h) - f(x)}{h}$$

and $f'(x)$ is read as f prime of x. The derivative function of f is a function that determines the slope of the tangent lines to a graph of f. Similarly, $f(a)$ is defined as

$$f'(a) = \lim_{h \to 0} \frac{f(a+h) - f(a)}{h}$$

and $f'(a)$ determines the slope of the tangent to the graph of f at $x = a$.

 The common notations for the derivative are $f'(x)$, $\frac{dy}{dx}$, y', $\frac{d}{dx} f(x)$.

Example 1   Finding the derivative function

Find the derivative function of $f(x) = x^3$ and evaluate $f'(-1)$.

Solution   Let's find $f'(x)$ using the definition of the derivative function.

$$\begin{aligned}
f'(x) &= \lim_{h \to 0} \frac{f(x+h) - f(x)}{h} \\
&= \lim_{h \to 0} \frac{(x+h)^3 - x^3}{h} \\
&= \lim_{h \to 0} \frac{x^3 + 3x^2h + 3xh^2 + h^3 - x^3}{h} \\
&= \lim_{h \to 0} \frac{h(3x^2 + 3xh + h^2)}{h} \\
&= \lim_{h \to 0} 3x^2 + 3xh + h^2 \\
&= 3x^2
\end{aligned}$$

Since $f'(x) = 3x^2$, $f'(-1) = 3(-1)^2 = 3$.

12.2 Finding the derivative functions

> ### Differentiation
>
> Differentiation means finding the derivative function. In general, finding the derivative function using the definition of the derivative is very tedious and takes long time. In this lesson, you will learn about the differentiation rules without using the definition of the derivative so that you can find the derivative function with ease.
>
> The table below summarizes the basic differentiation rules.
>
> | Basic differentiation rules | Example |
> |---|---|
> | 1. $\frac{d}{dx}(c) = 0$ | 1. $\frac{d}{dx}(2) = 0$ |
> | 2. $\frac{d}{dx}cf(x) = c \cdot \frac{d}{dx}f(x)$ | 2. $\frac{d}{dx}2x^2 = 2 \cdot \frac{d}{dx}x^2$ |
> | 3. $\frac{d}{dx}x^n = nx^{n-1}$ | 3. $\frac{d}{dx}x^3 = 3x^{3-1} = 3x^2$ |
> | 4. $\frac{d}{dx}[f(x) \pm g(x)] = \frac{d}{dx}f(x) \pm \frac{d}{dx}g(x)$ | 4. $\frac{d}{dx}(x^3 + x) = \frac{d}{dx}x^3 + \frac{d}{dx}x = 3x^2 + 1$ |
> | 5. $\frac{d}{dx}a^x = a^x \cdot \ln a$ | 5. $\frac{d}{dx}e^x = e^x \cdot \ln e = e^x$ |
> | 6. $\frac{d}{dx}\log_a x = \frac{1}{x \cdot \ln a}$ | 6. $\frac{d}{dx}\ln x = \frac{1}{x \cdot \ln e} = \frac{1}{x}$ |

Tip 1. Note that $\frac{d}{dx}cf(x) \neq \frac{d}{dx}c \cdot \frac{d}{dx}f(x)$. For instance,

$$\frac{d}{dx}2x^2 \neq \frac{d}{dx}2 \cdot \frac{d}{dx}x^2$$

2. $\frac{d}{dx}x^n = nx^{n-1}$ is called the **Power Rule**. You can apply the power rule to any power function $f(x) = x^n$, where the base is variable and exponent is constant. The derivative function of the following power functions are worth memorizing.

$$\frac{d}{dx}x = 1, \quad \frac{d}{dx}x^2 = 2x, \quad \frac{d}{dx}x^3 = 3x^2, \quad \frac{d}{dx}\sqrt{x} = \frac{1}{2\sqrt{x}}, \quad \frac{d}{dx}\frac{1}{x} = -\frac{1}{x^2}$$

Example 2   Finding the derivative function using the power rule

Find the derivative function of $f(x) = x^3$ using the power rule.

Solution

$$f'(x) = 3x^{3-1} = 3x^2$$

Example 3   Applying the differentiation rules

Differentiate $y = (x-1)^2$.

Solution   Since $(x-1)^2 = x^2 - 2x + 1$, differentiate each term.

$$\begin{aligned}
\frac{dy}{dx} &= (x^2 - 2x + 1)' \\
&= (x^2)' - 2(x)' + (1)' \\
&= 2x - 2(1) + 0 \\
&= 2x - 2
\end{aligned}$$

Example 4   Applying the differentiation rules

Differentiate $f(x) = \sqrt{x} + \dfrac{2}{x}$.

Solution   Since $\sqrt{x} = x^{\frac{1}{2}}$ and $\dfrac{1}{x} = x^{-1}$, apply the power rule.

$$\begin{aligned}
f'(x) &= (\sqrt{x})' + \left(\frac{2}{x}\right)' \\
&= (x^{\frac{1}{2}})' + 2(x^{-1})' \\
&= \frac{1}{2}x^{\frac{1}{2}-1} + 2(-1)x^{-1-1} \\
&= \frac{1}{2}x^{-\frac{1}{2}} - 2x^{-2} \\
&= \frac{1}{2\sqrt{x}} - \frac{2}{x^2}
\end{aligned}$$

12.3 Tangent and normal lines

Tangent and Normal Lines

The **tangent line** to the curve $y = f(x)$ at the point $A(x_1, y_1)$ is the line with the slope $m = f'(x_1)$.

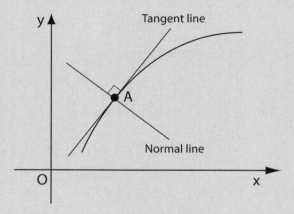

The equation of the tangent line with the slope $f'(x_1)$ at the point $A(x_1, y_1)$ using the point-slope form is given by

$$y - y_1 = f'(x_1)(x - x_1)$$

The **normal line** at the point $A(x_1, y_1)$ is perpendicular to the tangent line. So the slope of the normal line is $-\dfrac{1}{f'(x_1)}$. The equation of the normal line is given by

$$y - y_1 = -\frac{1}{f'(x_1)}(x - x_1), \qquad m \neq 0$$

Tip The point-slope form of a line through the point (x_1, y_1) with the slope m is given by

$$y - y_1 = m(x - x_1)$$

Example 5   Finding equations of the tangent and normal lines

Find equations of the tangent and normal lines to $y = 2x^3 - 4x^2 + 1$ at $x = 2$.

Solution   At $x = 2$, $y = 2(2)^3 - 4(2)^2 + 1 = 1$. So the point of tangency is $(2, 1)$.

$$\frac{dy}{dx} = 6x^2 - 8x$$

The slope of the tangent line at $x = 2$ is

$$\left.\frac{dy}{dx}\right|_{x=2} = 6(2)^2 - 8(2) = 8$$

Thus, the equation of the tangent line with the slope 8 at the point $(2, 1)$ is

$$y - 1 = 8(x - 2)$$
$$y = 8x - 15$$

The slope of the normal line is $-\dfrac{1}{8}$. Thus, the equation of the normal line with the slope $-\dfrac{1}{8}$ at the point $(2, 1)$ is

$$y - 1 = -\frac{1}{8}(x - 2)$$
$$y = -\frac{1}{8}x + \frac{5}{4}$$

EXERCISES

1. Differentiate with respect to x.

   (a) x^5

   (b) $\sqrt[3]{x}$

   (c) $x\sqrt{x}$

   (d) $\dfrac{x^3\sqrt{x}+1}{x^2}$

   (e) $\dfrac{1}{\sqrt[3]{x^2}}$

2. Differentiate with respect to x.

(a) $5x^4 + 4x^3 + 3x^2 + x + 1$

(b) $x - \dfrac{1}{x} - \dfrac{1}{x^2}$

(c) $(x - 1)^3$

(d) $(2x^2 - 1)(x^3 + 1)$

(e) $\dfrac{x^2 - x + 1}{\sqrt{x}}$

3. Find the value of $\dfrac{dy}{dx}$ at the given point on the curve.

    (a) $y = 4x^3 + 2x - 1$ at the point $(1, 1)$.

    (b) $y = 1 + \dfrac{2}{x^2}$ at the point $\left(\dfrac{1}{2}, 9\right)$.

    (c) $y = \dfrac{x - 6}{\sqrt{x}}$ at the point $(9, 1)$.

    (d) $y = \dfrac{3x^2 - 5\sqrt{x}}{x}$ at the point $(1, -2)$.

4. The slope of the tangent line to the curve $y = ax + \dfrac{2b}{x}$ at the point $(5, -6)$ is $-\dfrac{4}{5}$. Find the values of a and b.

5. $y = \dfrac{2}{3}x^3 - \dfrac{5}{2}x^2 + 2x$.

   (a) Find $\dfrac{dy}{dx}$

   (b) Find the range of values of x for which $\dfrac{dy}{dx} \leq 0$

6. The tangent lines to the curve $y = x^3 - 3x^2 - 4x + 12$ at the points $A(-1, 12)$ and $B(1, 6)$ intersect at the point C.

   (a) Find the coordinates of C.

   (b) Find the area of triangle ABC.

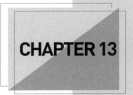 

CHAPTER 13 | Differentiation rules

13.1 The product and quotient rules

The Product and Quotient Rules

Product rule and quotient rule are the differentiation rules that enable us to find the derivative of product function and quotient function.

- The Product Rule: If f and g are both differentiable, then

$$(f \cdot g)' = f' \cdot g + f \cdot g'$$

- The Quotient Rule: If If f and g are both differentiable, then

$$\left(\frac{f}{g}\right)' = \frac{f' \cdot g - f \cdot g'}{g^2}$$

---

 1. Note that the derivative of a product of two functions is **NOT** the product of derivatives as shown below.

$$(f \cdot g)' \neq f' \cdot g'$$

2. Note that the derivative of a quotient of two functions is **NOT** the quotient of derivatives as shown below.

$$\left(\frac{f}{g}\right)' \neq \frac{f'}{g'}$$

Example 1   Applying the Product rule

If $f(x) = xe^x$, find $f'(x)$.

Solution

$$f'(x) = (xe^x)' = (x)' \cdot e^x + x \cdot (e^x)'$$
$$= 1 \cdot e^x + xe^x = e^x + xe^x$$

Derivatives of Six Trigonometric Functions

$$\frac{d}{dx}(\sin x) = \cos x \qquad \frac{d}{dx}(\cos x) = -\sin x$$

$$\frac{d}{dx}(\tan x) = \sec^2 x \qquad \frac{d}{dx}(\cot x) = -\csc^2 x$$

$$\frac{d}{dx}(\sec x) = \sec x \tan x \qquad \frac{d}{dx}(\csc x) = -\csc x \cot x$$

Tip Note that all the derivatives of the trigonometric functions starting with the letter C have negative signs.

Example 2   Applying the Quotient rule

Differentiate $y = \tan x = \dfrac{\sin x}{\cos x}$ using the Quotient rule.

Solution

$$\frac{dy}{dx} = \left(\frac{\sin x}{\cos x}\right)' \qquad \text{Apply the quotient rule}$$

$$= \frac{(\sin x)' \cdot \cos x - \sin x \cdot (\cos x)'}{\cos^2 x}$$

$$= \frac{\cos x \cos x - \sin x \cdot (-\sin x)}{\cos^2 x}$$

$$= \frac{\cos^2 x + \sin^2 x}{\cos^2 x} \qquad \text{Use } \cos^2 x + \sin^2 x = 1$$

$$= \frac{1}{\cos^2 x}$$

$$= \sec^2 x$$

Example 3   Applying the Product rule

Find the derivative function of $y = e^x \sin x$.

Solution   Since $e^x \sin x$ is a product of two functions, apply the Product rule to the function.

$$\frac{dy}{dx} = (e^x \sin x)'$$

$$= (e^x)' \cdot \sin x + e^x \cdot (\sin x)' \qquad \text{Since } (e^x)' = e^x$$

$$= e^x \sin x + e^x \cos x$$

13.2 The chain rule

The Chain Rule

If you want to differentiate the function $f(x) = (x+1)^3$ using the differentiation rules that you have learned so far, you need to expand $(x+1)^3$ and apply the sum rule and power rule to each term of the function. Since $(x+1)^3 = x^3 + 3x^2 + 3x + 1$, the derivative function is

$$\frac{d}{dx}(x+1)^3 = (x^3 + 3x^2 + 3x + 1)' = 3x^2 + 6x + 3$$

However, a serious problem arises when you want to differentiate the function $(x+1)^{100}$. According to the Binomial theorem, $(x+1)^{100}$ can be written as

$$(x+1)^{100} = \sum_{k=0}^{100} \binom{100}{k} x^{100-k} 1^k$$

which indicates that the function $(x+1)^{100}$ has 101 terms. So, if you want to differentiate the function $f(x) = (x+1)^{100}$, you need to expand the function and differentiate 101 terms of the function. In this lesson, you will learn about the new differentiation rule called the **Chain Rule** which will help you differentiate a composition function like $f(x) = (x+1)^{100}$ with ease.

- The Chain Rule

  If f and g are both differentiable and F is the composition function defined by $F(x) = f(g(x))$, then F' is given by the product

  $$F'(x) = f'(g(x))g'(x)$$

  In Leibniz notion, if $y = f(u)$ and $y = g(x)$ are both differentiable function, then

  $$\frac{dy}{dx} = \frac{dy}{du} \cdot \frac{du}{dx}$$

Tip   The Chain Rule is the one of the most important differentiation rules that you are going to use a lot throughout the AP Calculus AB and AP Calculus BC courses. Many students tend to make mistake by forgetting the Chain rule when they differentiate a composition function.

Example 4　Applying the Chain rule

Differentiate $y = (x+1)^{100}$ using the Chain rule.

Solution　Let $y = u^{100}$ and $u = x+1$. If you substitute u for $x+1$, then $y = (x+1)^{100}$. Since $(x+1)^{100}$ is a composition function, apply the Chain rule to the function.

$$y = u^{100}$$
$$\frac{dy}{du} = 100u^{99}$$
$$= 100(x+1)^{99}$$

$$u = x+1$$
$$\frac{du}{dx} = 1$$

Thus,

$$\frac{dy}{dx} = \frac{dy}{du} \cdot \frac{du}{dx}$$
$$= 100(x+1)^{99} \cdot 1$$
$$= 100(x+1)^{99}$$

Therefore, the derivative function of $y = (x+1)^{100}$ is $100(x+1)^{99}$.

Example 5　Applying the Chain rule

Differentiate $y = \sqrt{1-x^2}$.

Solution　Let $y = \sqrt{u}$ and $u = 1-x^2$. If you substitute u for $1-x^2$, then $y = \sqrt{1-x^2}$. Since $y = \sqrt{1-x^2}$ is a composition function, apply the Chain rule to the function.

$$y = \sqrt{u}$$
$$\frac{dy}{du} = \frac{1}{2\sqrt{u}}$$
$$= \frac{1}{2\sqrt{1-x^2}}$$

$$u = 1-x^2$$
$$\frac{du}{dx} = -2x$$

Thus,

$$\begin{aligned}
\frac{dy}{dx} &= \frac{dy}{du} \cdot \frac{du}{dx} \\
&= \frac{1}{2\sqrt{1-x^2}} \cdot -2x \\
&= \frac{-x}{\sqrt{1-x^2}}
\end{aligned}$$

Therefore, the derivative function of $y = \sqrt{1-x^2}$ is $\dfrac{-x}{\sqrt{1-x^2}}$.

Example 6   Applying the Chain rule

Differentiate $y = e^{\sin x}$.

Solution   Since $e^{\sin x}$ is a composition function, let $y = e^u$ and $u = \sin x$. Use the Chain rule to differentiate the function.

$$
\begin{array}{l|l}
y = e^u & u = \sin x \\[2mm]
\dfrac{dy}{du} = e^u & \dfrac{du}{dx} = \cos x \\[2mm]
\quad\ = e^{\sin x} &
\end{array}
$$

Thus,

$$\begin{aligned}
\frac{dy}{dx} &= \frac{dy}{du} \cdot \frac{du}{dx} \\
&= e^{\sin x} \cdot \cos x \\
&= e^{\sin x} \cos x
\end{aligned}$$

Therefore, the derivative function of $y = e^{\sin x}$ is $e^{\sin x} \cos x$.

13.3 The second derivative

The Second Derivative

If you differentiate $y = f(x)$ with respect to x, you obtain the **first derivative** which is denoted by

$$f'(x) = y' = \frac{dy}{dx}$$

If you differentiate the first derivative with respect to x, you obtain the **second derivative** which is denoted by

$$f''(x) = y'' = \frac{d^2y}{dx^2}$$

Example 7   Finding the second derivative

Suppose $y = 3x^2 - \dfrac{2}{x^2}$.

(a) Find $\dfrac{d^2y}{dx^2}$.

(b) Find the value of $\dfrac{d^2y}{dx^2}$ when $x = -1$.

Solution

(a) Rewrite $y = 3x^2 - \dfrac{2}{x^2}$ as $y = 3x^2 - 2x^{-2}$

$$\frac{dy}{dx} = 6x + 4x^{-3}$$

$$\frac{d^2y}{dx^2} = 6 - 12x^{-4} = 6 - \frac{12}{x^4}$$

(b)

$$\left. \frac{d^2y}{dx^2} \right|_{x=-1} = 6 - \frac{12}{(-1)^4} = -6$$

General forms of the derivatives

$$\frac{d}{dx}\Big[\sin(ax+b)\Big] = a\cos(ax+b)$$

$$\frac{d}{dx}\Big[\cos(ax+b)\Big] = -a\sin(ax+b)$$

$$\frac{d}{dx}\Big[\tan(ax+b)\Big] = a\sec^2(ax+b)$$

$$\frac{d}{dx}\Big[e^{ax+b}\Big] = a \times e^{ax+b}$$

$$\frac{d}{dx}\Big[\ln(ax+b)\Big] = \frac{a}{ax+b}$$

$$\frac{d}{dx}\Big[e^{f(x)}\Big] = f'(x) \times e^{f(x)}$$

$$\frac{d}{dx}\Big[\ln f(x)\Big] = \frac{f'(x)}{f(x)}$$

EXERCISES

1. Differentiate the following function with respect to x.

   (a) $y = (2x^2 + 3x)^{10}$

   (b) $y = \left(\dfrac{x+1}{x-1}\right)^{10}$

   (c) $y = \ln \sqrt{x}$

   (d) $y = \sin^2 x$

2. Find the gradient of the tangent to the following curve at the given point.

(a) $y = \cos 4\theta$ at the point $\theta = \dfrac{\pi}{3}$

(b) $y = \dfrac{4}{\sqrt{2x-3}}$ at the point $x = 6$

(c) $y = \sqrt[3]{1 + \sin\theta}$ at the point $\theta = \pi$

(d) $y = \ln(x^2 + 1)$ at the point $x = 2$

3. Find $\dfrac{dy}{dx}$.

   (a) $y = e^{2x-1}$

   (b) $y = x^2 \sin 2x$

   (c) $y = \dfrac{\ln x}{2x-1}$

   (d) $e^y = \cos 4x$

   (e) $y = \ln\left[\dfrac{x-1}{(x+1)(x-2)}\right]$

4. Find $\dfrac{d^2y}{dx^2}$.

    (a) $y = 4x^3 - 12x^2 + 6x - 5$

    (b) $y = \sqrt{3x - 2}$

    (c) $y = \dfrac{x - 2}{x + 3}$

    (d) $y = x^2 e^{3x}$

5. A curve which has equation $y = \dfrac{x-2}{\sqrt{x+4}}$ intersects the line $y = 1$ at point A. Find the equation of the tangent and the normal to the curve $y = \dfrac{x-2}{\sqrt{x+4}}$ at point A.

6. A curve has equation $y = \dfrac{4}{3}x^3 - 10x^2 + 16x + 1$. Complete the table below to show whether $\dfrac{dy}{dx}$ and $\dfrac{d^2y}{dx^2}$ are positive, negative, or zero for the given values of x.

x	0	1	2	3	4	5
$\dfrac{dy}{dx}$						
$\dfrac{d^2y}{dx^2}$						

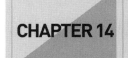 

CHAPTER 14 Applications of differentiation

14.1 Small increments and approximations

Small Increments and Approximations

The figure shows the tangent to the curve $y = f(x)$ at the point P.

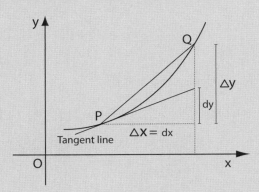

The gradient of the tangent line at the point P is $\dfrac{dy}{dx}$. The gradient of the chord PQ is $\dfrac{\Delta y}{\Delta x}$.
If the points P and Q are very close then

$$\frac{\Delta y}{\Delta x} \approx \frac{dy}{dx} \qquad \text{or} \qquad \Delta y \approx \frac{dy}{dx} \cdot \Delta x$$

Rearranging $\dfrac{\Delta y}{\Delta x} \approx \dfrac{dy}{dx}$ to get Δx gives,

$$\Delta x \approx \frac{dx}{dy} \cdot \Delta y, \qquad \text{where} \quad \frac{dx}{dy} = \frac{1}{\frac{dy}{dx}}$$

1. The notation $\dfrac{\Delta y}{\Delta x}$ can be denoted by $\dfrac{\delta y}{\delta x}$.

2. From $\Delta y \approx \dfrac{dy}{dx} \cdot \Delta x$, the expression $\dfrac{dy}{dx} \cdot \Delta x$ is referred to as the approximate change in y.

Example 1   Finding the approximate change

Variables x and y are related by the equation $y = x^3 - 2x^2$.

(a) Find the approximate change in y as x increases from 3 to 3.01.

(b) Find the approximate change in x as y increases from 9 to 9.05.

Solution

(a) $y = x^3 - 2x^2$. So $\dfrac{dy}{dx} = 3x^2 - 4x$. When $x = 3$, the value of $\dfrac{dy}{dx} = 3(3)^2 - 4(3) = 15$. Since x increases from 3 to 3.01, $\Delta x = 0.01$. Thus,

$$\frac{\Delta y}{\Delta x} \approx \frac{dy}{dx}$$
$$\Delta y \approx \frac{dy}{dx} \cdot \Delta x$$
$$\Delta y \approx 15(0.01)$$
$$\Delta y \approx 0.15$$

Therefore, the approximate change in y is 0.15.

(b) $\dfrac{dx}{dy} = \dfrac{1}{\frac{dy}{dx}} = \dfrac{1}{3x^2 - 4x}$, and $\Delta y = 0.05 = \dfrac{1}{20}$. When $y = 9$, $x = 3$.

$$\left.\frac{dx}{dy}\right|_{x=3} = \frac{1}{3(3)^2 - 4(3)} = \frac{1}{15}$$

Thus,

$$\Delta x \approx \frac{dx}{dy} \cdot \Delta y = \frac{1}{15} \times \frac{1}{20} = \frac{1}{300}$$

Therefore, the approximate change in x is $\dfrac{1}{300}$.

Example 2   Finding the approximate change

The volume, V cm^3, of a sphere with radius r cm is $V = \dfrac{4}{3}\pi r^3$. Find the approximate change in V as r increases from 2 to $2 + \delta r$.

Solution $V = \dfrac{4}{3}\pi r^3$. So $\dfrac{dV}{dr} = 4\pi r^2$. When $r = 2$, the value of $\dfrac{dV}{dr} = 4\pi(2)^2 = 16\pi$. Since r increases from 2 to $2 + \delta r$, $\Delta r = \delta r$. Thus,

$$\frac{\Delta V}{\Delta r} \approx \frac{dV}{dr}$$
$$\Delta V \approx \frac{dV}{dr} \cdot \Delta r$$
$$\Delta V \approx 16\pi \delta r$$

Therefore, the approximate change in V, is $16\pi \delta r$.

14.2 Related rates

Related Rates

In a related rates problem, we are trying to find the rate of change of one quantity in terms of the rate of change of another quantity that is already measured.

When two variables x and y are both functions of a third variable t, the three variables can be connected using the chain rule.

$$\frac{dy}{dt} = \frac{dy}{dx} \times \frac{dx}{dt}$$

Example 3   Solving a related rates problem

A ladder 10 meter-long leans against a building. If the bottom of the ladder slides away from the wall at a rate of 2 ms^{-1}, how fast is the top of the ladder sliding down the wall when the bottom of the ladder is 6 meters from the wall?

Solution   Let x be the distance from the bottom of the ladder to the wall and y be the distance from the top of the ladder to the ground as shown in figures below. Since the top of the ladder slides down, y decreases as time increases. However, the bottom of the lader slides away. So, x increases as time increases. Thus, x and y are functions of time t.

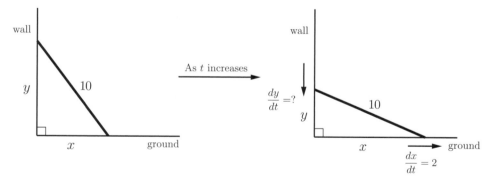

We are given that $\dfrac{dx}{dt} = 2$ ms^{-1} and try to find $\dfrac{dy}{dt}$ when $x = 6$ meters. Let's set up an equation that relates x and y using the Pythagorean theorem.

$$x^2 + y^2 = 10^2$$
$$y = \sqrt{100 - x^2}, \qquad \text{Since } y > 0$$

Differentiating y with respect to x using the chain rule gives

$$\frac{dy}{dx} = \frac{1}{2\sqrt{100 - x^2}} \cdot (-2x)$$
$$= \frac{-x}{\sqrt{100 - x^2}}$$

When $x = 6$, the value of $\frac{dy}{dx}$ is

$$\left.\frac{dy}{dx}\right|_{x=6} = \frac{-6}{\sqrt{100 - 6^2}} = -\frac{3}{4}$$

Using the chain rule,

$$\frac{dy}{dt} = \frac{dy}{dx} \times \frac{dx}{dt}$$
$$= -\frac{3}{4} \times 2$$
$$= -\frac{3}{2}$$

Therefore, the top of the ladder is sliding down the wall at a rate of $-\frac{3}{2}$ ms^{-1}.

Example 4 Solving a related rates problem

If a spherical snowball melts so that its volume decreases at a rate of $1 \text{ cm}^3/\text{min}$, find the rate at which the radius decreases when the radius is 10 cm.

Solution   Let V be the volume of the sphere and r be the radius of the sphere. As shown in the figures below, both the volume and radius decrease as time t increases. So, the volume and the radius are the functions of t.

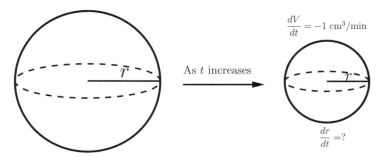

We are given that $\dfrac{dV}{dt} = -1 \text{ cm}^3/\text{min}$ and try to find $\dfrac{dr}{dt}$ when $r = 10$ cm. Let's set up an equation that relates V and r.

$$V = \frac{4}{3}\pi r^3$$

Differentiating V with respect to r using the chain rule gives

$$\frac{dV}{dr} = 4\pi r^2$$

When $r = 10$, the value of $\dfrac{dV}{dr}$ is

$$\left.\frac{dV}{dr}\right|_{r=10} = 4\pi(10)^2 = 400\pi$$

Using the chain rule,

$$\begin{aligned}
\frac{dr}{dt} &= \frac{dr}{dV} \times \frac{dV}{dt} \\
&= \frac{1}{400\pi} \times -1 \\
&= -\frac{1}{400\pi}
\end{aligned}$$

Therefore, the rate at which the radius decreases when the radius is 10 cm is $-\dfrac{1}{400\pi} \text{ cm/min}$.

14.3 Understanding a curve from the first and second derivatives

---

Increasing and Decreasing Test

Increasing and Decreasing Test

1. If $f'(x) > 0$ for all x on an interval, then f is increasing on that interval.

2. If $f'(x) < 0$ for all x on an interval, then f is decreasing on that interval.

Example 5   Finding where a function is increasing or decreasing

Find where the function $f(x) = x^3 - 9x^2 + 24x + 1$ is increasing and where it is decreasing.

Solution $f'(x) = 3x^2 - 18x + 24 = 3(x - 2)(x - 4)$.

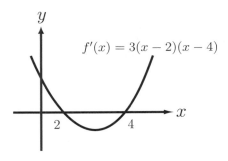

As shown in the figure above, $f'(x) > 0$ for $x < 2$ and also for $x > 4$. Whereas, $f'(x) < 0$ for $2 < x < 4$. Therefore, f is increasing for $x < 2$ and $x > 4$, and f is decreasing for $2 < x < 4$.

Concavity

If the graph of f lies above all of its tangents on an interval, then it is called **concave upward** on that interval. Whereas, if the graph of f lies below all of its tangent on an interval, it is called **concave downward** on that interval.

Concavity Test

Concavity Test

1. If $f''(x) > 0$ for all x on an interval, then the graph of f is concave upward on that interval.

2. If $f''(x) < 0$ for all x on an interval, then the graph of f is concave downward on that interval.

 Tip
1. $f''(x) > 0$ means that $f'(x)$ is increasing. Thus the graph of f is concave upward.

2. $f''(x) < 0$ means that $f'(x)$ is decreasing. Thus the graph of f is concave downward.

Inflection Point

A point P on the curve is called an **inflection point** if the curve changes from concave upward to concave downward or from concave downward to concave upward at P.

If an inflection point on the curve at $x = a$, $f''(a) = 0$.

Example 6   Finding the intervals of concavity and the inflection point

If $f(x) = x^3 - 9x^2 + 24x + 1$, find the intervals of concavity and the inflection point.

Solution $f'(x) = 3x^2 - 18x + 24 = 3(x-2)(x-4)$ and $f''(x) = 6x - 18$. Since $f''(x) = 0$ when $x = 3$, $(3, 19)$ is the inflection point. Using the Concavity test,

$$f''(x) > 0 \quad \text{for } x > 3 \quad \implies \quad f \text{ is concave upward}$$
$$f''(x) < 0 \quad \text{for } x < 3 \quad \implies \quad f \text{ is concave downward}$$

Therefore, f is concave upward for $x > 3$, and f is concave downward for $x < 3$.

14.4 Local maximum and local minimum

Definition of a Stationary point

A **stationary point**, **critical number**, **turning point** of a function f is a number c in the domain of f such that $f'(c) = 0$.

 Tip   The stationary point is important when you find a local maximum or a local minimum because f might have a local maximum or a local minimum at the stationary point.

Example 7   Finding stationary points

If $f(x) = x^3 - 3x^2 - 9x$, find the stationary points of f.

Solution   Find the derivative of f. So, $f'(x) = 3x^2 - 6x - 9$. Set the derivative of f equal to 0 and solve for x.

$$3x^2 - 6x - 9 = 0$$
$$3(x^2 - 2x - 3) = 0$$
$$3(x + 1)(x - 3) = 0$$
$$x = -1 \quad \text{or} \quad x = 3$$

Therefore, the stationary points of f are -1 and 3.

Local Maximum and Local Minimum

In the figure, the stationary point P is called a **local maximum** (or relative maximum) because the value of y at this point is greater than the value of y at other points close to P.

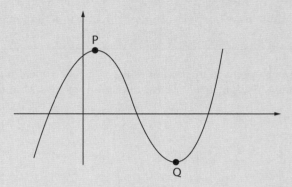

The stationary point Q is called a **local minimum** (or relative minimum) because the value of y at this point is less than the value of y at other points close to Q.

- At the local maximum point (at the point P): $f'(x) = 0$.
- At the local minimum point (at the point Q): $f'(x) = 0$.

Finding a Local Maximum and a Local Minimum

There are two ways to find a local maximum and a local minimum: **The first derivative test** and **the second derivative test**.

The First Derivative Test: Suppose that c is a stationary point of a continuous function f.

1. If f' changes from positive to negative at c, then f has a local maximum at c.

2. If f' changes from negative to positive at c, then f has a local minimum at c.

The Second Derivative Test: Suppose f'' is continuous near c.

1. If $f'(c) = 0$ and $f''(c) > 0$, then f has a local minimum at c.

2. If $f'(c) = 0$ and $f''(c) < 0$, then f has a local maximum at c.

Example 8 Finding the local maximum and local minimum

If $f(x) = x^3 - 9x^2 + 24x + 1$, find the local maximum and local minimum values of f.

Solution $f'(x) = 3x^2 - 18x + 24 = 3(x-2)(x-4)$ and $f''(x) = 6x - 18$. $f'(x) = 0$ for $x = 2$ and $x = 4$. So, the stationary points are 2 and 4. Let's find the local maximum and local minimum values of f using both the First Derivative test and Second Derivative test.

- Using the First Derivative Test:

  As shown in the figure below, f' changes positive to negative at $x = 2$. Thus, f has a local maximum at $x = 2$. The local maximum of value of f is $f(2) = 21$.

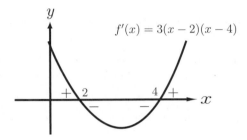

  However, f' changes negative to positive at $x = 4$. Thus, f has a local minimum at $x = 4$. The local minimum of value of f is $f(4) = 17$.

- Using the Second Derivative Test:

  Since the stationary points are 2 and 4, substituting these values into the second derivative $f''(x) = 6x - 18$ will determine the local maximum and local minimum value of f.

  $$f''(2) = 6(2) - 18 < 0 \implies f \text{ has the local maximum at } x = 2$$
  $$f''(4) = 6(4) - 18 > 0 \implies f \text{ has the local minimum at } x = 4$$

  Therefore, f has the local maximum of value of f is $f(2) = 21$ and the local minimum of value of f is $f(4) = 17$.

14.5 Practical maximum and minimum problems

Optimization Problems

Optimization is one of the most important applications of the first derivative because it has many applications in real life. In general, optimization consists of maximizing or minimizing a function with a constraint. The following guidelines will help you solve optimization problems.

Guidelines for solving maximum and minimum problems

1. Read the problem and draw a diagram.

2. Define variables and label your diagram with these variables. It will help you set up mathematical equations.

3. Set up two equations that are related to the diagram. One equation is an optimization equation and the other is a constraint equation. The constraint equation is used to solve for one of the variables.

4. Set up the optimization equation as a function of only one variable using the constraint equation.

5. Differentiate the optimization equation and find the stationary points.

6. Find the local maximum and minimum using the first or second derivative tests.

Example 9   Solving a minimum problem

Find the two positive numbers whose product is 100 and whose sum is minimum.

Solution   Let x and y be the two positive numbers. So, the sum S and constraint can be defined as

$$\text{Minimization equation:} \quad S(x,y) = x + y$$
$$\text{Constraint equation:} \quad xy = 100$$

From the constraint equation, we get $y = \dfrac{100}{x}$. Substituting this into the minimization equation,

$$S(x) = x + \frac{100}{x}$$

Differentiate $S(x)$ with respect to x and find the stationary points.

$$S'(x) = 1 - \frac{100}{x^2} = \frac{x^2 - 100}{x^2}$$

Since x is the positive number, $S'(x) = 0$ when $x = 10$. So, the stationary point is 10. Using the First Derivative test, $S'(x) < 0$ when $x < 10$ and $S(x) > 0$ when $x > 10$. Thus, the function $S(x)$ has the local minimum at $x = 10$. Therefore, the sum S is minimum when $x = 10$ and $y = \dfrac{100}{x} = 10$.

Example 10   Solving a maximum problem

A farmer has 1000 m of fencing and wants to fence off a rectangular field that borders a straight river. He does not need any fence along the river. Find the dimensions of the field that has the largest area.

Solution   As shown in the figure below, let x and y be the width and length of the rectangular field, respectively.

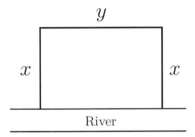

So, the area A and constraint can be defined as

$$\text{Maximization equation:} \quad A(x, y) = xy$$
$$\text{Constraint equation:} \quad 2x + y = 1000$$

From the constraint equation, we get $y = 1000 - 2x$. Substituting this into the maximization equation,

$$A(x) = x(1000 - 2x) = 1000x - 2x^2$$

Differentiate $A(x)$ with respect to x and find the stationary points.

$$A'(x) = 1000 - 4x$$

Since $A'(x) = 0$ when $x = 250$, the stationary point is 250. Note that $A''(x) = -4$ for all x. Using the Second Derivative test, $A''(250) < 0$ which indicates that the function $A(x)$ has the local maximum at $x = 250$. Thus, $y = 1000 - 2x = 1000 - 2(250) = 500$. Therefore, the dimensions of the field that has the largest area are width of 250 m and length of 500 m.

EXERCISES

1. Variables x and y are related by the equation $y = x^2 \sqrt{x}$.

   (a) Find $\dfrac{dy}{dx}$.

   (b) Find $\dfrac{dx}{dy}$ when $y = 1$.

   (c) Find the approximate change in y as x increases from 1 to 1.1.

   (d) Find the approximate change in x as y increases from 32 to 32.025.

2. The side length x cm of a cube is increasing at a rate of 2 cm s^{-1}.

   (a) Find the rate at which the volume V cm^3 of the cube is increasing when $x = 5$.

   (b) Find the rate at which the surface area A cm^2 of the cube is increasing when $x = 5$.

   (c) Find the approximate change in V as x increases from 10 to 10.2.

   (d) Find the approximate change in A as x increases from 10 to 10.2.

3. A vase with a circular cross-section is shown in the figure.

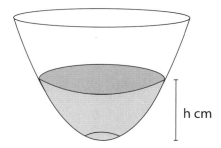

Water is poured into the vase at a rate of 60π cm^3 s^{-1}. When the water level is h cm, the volume of the water, in V cm^3, in the vase is given by

$$V = 5\pi h(h + 6), \qquad 0 \le h \le 30$$

(a) Find the rate at which the water level is rising, in cm s^{-1}, when $h = 4$ cm.

(b) Find the approximate change in V as h increases from 6 cm to $6 + \dfrac{1}{15}$ cm.

4. A conical tank with height x cm is shown in the figure. The volume of a cone is $V = \frac{1}{3}\pi r^2 h$, where r is the radius of the base and h is the height of the cone.

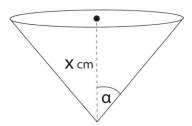

Initially, the tank is empty. Then water is poured into the tank at a rate of $10\,\text{cm}^3\,\text{s}^{-1}$. The depth of the water at time t seconds is x cm. The angle between the axis and the slant height of the conical tank is α, where $\alpha = \tan^{-1}\left(\frac{2}{3}\right)$.

(a) Show that the volume, V cm^3, of water in the tank when the depth is x cm is given by

$$V = \frac{4}{27}\pi x^3$$

(b) The rate at which the depth of the water is increasing, in cm s^{-1}, when the depth is 9 cm can be expressed as $\frac{n}{\pi}$ cm s^{-1}. Find the value of n.

5. Find the interval of increase, and local maximum and minimum values of the following functions.

(a) $f(x) = x^3 - 3x^2 + 3$

(b) $f(x) = x^2 e^x$

(c) $f(x) = x - \sqrt{x - 1}$

(d) $f(x) = \dfrac{x}{x^2 + 1}$

6. Find the dimensions of the rectangle of largest area that has its base on the x-axis and its other two vertices above the x-axis lying on the parabola $y = 12 - x^2$.

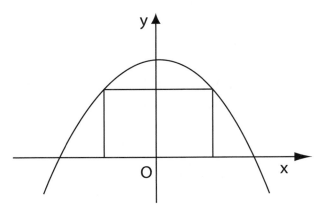

7. The figure shows a playground that consists of two semi-circular areas with centers P and Q, each of radius x meters.

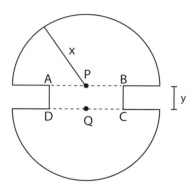

The two semi-circular areas are attached to each other by a rectangular area $ABCD$ with $AB = x$ meters and $BC = y$ meters. The area of the playground is 300 m^2.

(a) Show that the perimeter P meters of the playground is given by

$$P = 2x + \frac{600}{x}$$

(b) Find the minimum value of P.

CHAPTER 15 Integration

15.1 Indefinite integrals

Antiderivative or Indefinite Integral

The derivatives of four functions below are the same: $3x^2$.

$$(x^3 + 3)' = 3x^2, \qquad (x^3 + 2)' = 3x^2, \qquad (x^3 + 1)' = 3x^2, \qquad (x^3 + C)' = 3x^2$$

An antiderivative or indefinite integral of f is a differentiable function F such that the derivative of F is f; that is, $F' = f$. The process of solving for antiderivatives is called **indefinite integration** or **integration** and its opposite operation is differentiation. For instance, the antiderivative of $3x^2$ is $x^3 + C$, where C is a constant. This can be written as

$$\int 3x^2 dx = x^3 + C$$

In general,

$$\int f dx = F + C \qquad \text{because} \qquad \frac{d}{dx}(F + C) = f$$

Example 1   Finding the antiderivative

Find the antiderivative of $\cos x$.

Solution   Since $(\sin x + C)' = \cos x$,

$$\int \cos dx = \sin x + C$$

Thus, the antiderivative of $\cos x$ is $\sin x + C$.

Example 2   Finding the antiderivative

Find the indefinite integral of $x^3 + 4x + 1$.

Solution   Since $\left(\dfrac{1}{4}x^4 + 2x^2 + x + C\right)' = x^3 + 4x + 1$,

$$\int x^3 + 4x + 1 \, dx = \frac{1}{4}x^4 + 2x^2 + x + C$$

Basic Indefinite Integrals

1. $\displaystyle\int cf(x)\,dx = c\int f(x)\,dx$

2. $\displaystyle\int \left[f(x) \pm g(x)\right]dx = \int f(x)\,dx \pm \int g(x)\,dx$

3. $\displaystyle\int k\,dx = kx + C$

4. $\displaystyle\int x^n\,dx = \frac{1}{n+1}x^{n+1} + C \quad (n \neq -1)$

5. $\displaystyle\int \frac{1}{x}\,dx = \ln|x| + C$

6. $\displaystyle\int a^x\,dx = \frac{a^x}{\ln a} + C$

7. $\displaystyle\int e^x\,dx = e^x + C$

8. $\displaystyle\int \sin x\,dx = -\cos x + C$

9. $\displaystyle\int \cos x\,dx = \sin x + C$

10. $\displaystyle\int \sec^2 x\,dx = \tan x + C$

11. $\displaystyle\int \csc^2 x\,dx = -\cot x + C$

12. $\displaystyle\int \sec x \tan x\,dx = \sec x + C$

13. $\displaystyle\int \csc x \cot x\,dx = -\csc x + C$

Tip   Note that $\displaystyle\int f(x)\cdot g(x)dx \neq \int f(x)dx \cdot \int g(x)dx$

Example 3   Finding the indefinite integral

Find the indefinite integral of $f(x) = \sqrt[3]{x^2}$.

Solution   Since $\sqrt[3]{x^2}$ can be written as $x^{\frac{2}{3}}$,

$$\int \sqrt[3]{x^2}\,dx = \int x^{\frac{2}{3}}\,dx = \frac{1}{\frac{2}{3}+1}x^{\frac{2}{3}+1} + C$$
$$= \frac{3}{5}x^{\frac{5}{3}} + C$$

Example 4   Finding the indefinite integral

Find the indefinite integral of $f(x) = \dfrac{x^2 + x + 1}{x}$.

Solution

$$
\begin{aligned}
\int \frac{x^2 + x + 1}{x}\, dx &= \int \frac{x^2}{x} + \frac{x}{x} + \frac{1}{x}\, dx \\
&= \int x + 1 + \frac{1}{x}\, dx \\
&= \int x\, dx + \int 1\, dx + \int \frac{1}{x}\, dx \\
&= \frac{1}{2}x^2 + x + \ln|x| + C
\end{aligned}
$$

Example 5   Finding the indefinite integral

Given that $\dfrac{dy}{dx} = 4x^3 + \dfrac{2}{3x^5} - 4x$, find y in terms of x.

Solution

$$
\begin{aligned}
\frac{dy}{dx} &= 4x^3 + \frac{2}{3}x^{-5} - 4x \\
y &= 4\left(\frac{1}{4}\right)x^4 + \frac{2}{3}\left(-\frac{1}{4}\right)x^{-4} - 4\left(\frac{1}{2}\right)x^2 \\
&= x^4 - \frac{1}{6x^4} - 2x^2
\end{aligned}
$$

Example 6   Finding the equation of a curve

A curve is such that $\dfrac{dy}{dx} = \dfrac{x^3 - 1}{x^2}$ and the point $(1, 3)$ is on the curve. Find the equation of the curve.

Solution $\dfrac{dy}{dx} = \dfrac{x^3 - 1}{x^2} = x - \dfrac{1}{x^2}.$

$$\frac{dy}{dx} = x - \frac{1}{x^2}$$

$$y = \int x - x^{-2}\, dx$$

$$= \frac{1}{2}x^2 + \frac{1}{x} + C$$

The point $(1, 3)$ is on the curve. Substituting $x = 1$ and $y = 3$ into $y = \dfrac{1}{2}x^2 + \dfrac{1}{x} + C$ gives $C = \dfrac{3}{2}$.

Therefore, the equation of the curve is $y = \dfrac{1}{2}x^2 + \dfrac{1}{x} + \dfrac{3}{2}$.

Example 7   Understanding that integration is the reverse process of differentiation

Given that $y = x\sqrt{x^2 + 4}$.

(a) Find $\dfrac{dy}{dx}$.

(b) Hence find $\displaystyle\int \dfrac{x^2 + 2}{\sqrt{x^2 + 4}}\, dx$.

Solution

(a) $y = x\sqrt{x^2 + 4}$. Using the product rule,

$$\frac{dy}{dx} = 1 \cdot \sqrt{x^2 + 4} + x \cdot \frac{1}{2\sqrt{x^2 + 4}} \cdot 2x$$

$$= \sqrt{x^2 + 4} + \frac{x^2}{\sqrt{x^2 + 4}}$$

$$= \frac{(x^2 + 4) + x^2}{\sqrt{x^2 + 4}}$$

$$= \frac{2x^2 + 4}{\sqrt{x^2 + 4}}$$

(b) $\displaystyle\int \frac{x^2 + 2}{\sqrt{x^2 + 4}}\, dx = \frac{1}{2}\int \frac{2(x^2 + 2)}{\sqrt{x^2 + 4}}\, dx = \frac{1}{2}x\sqrt{x^2 + 4} + C.$

Tip   Note that $\displaystyle\int \frac{dy}{dx}\, dx = y + C$

15.2 The U-Substitution rule

Basic Indefinite Integrals

1. $\displaystyle\int cf(x)\,dx = c\int f(x)\,dx$

2. $\displaystyle\int \left[f(x) \pm g(x)\right]dx = \int f(x)\,dx \pm \int g(x)\,dx$

3. $\displaystyle\int k\,dx = kx + C$

4. $\displaystyle\int x^n\,dx = \frac{1}{n+1}x^{n+1} + C \quad (n \neq -1)$

5. $\displaystyle\int \frac{1}{x}\,dx = \ln|x| + C$

6. $\displaystyle\int a^x\,dx = \frac{a^x}{\ln a} + C$

7. $\displaystyle\int e^x\,dx = e^x + C$

8. $\displaystyle\int \sin x\,dx = -\cos x + C$

9. $\displaystyle\int \cos x\,dx = \sin x + C$

10. $\displaystyle\int \sec^2 x\,dx = \tan x + C$

11. $\displaystyle\int \csc^2 x\,dx = -\cot x + C$

12. $\displaystyle\int \sec x \tan x\,dx = \sec x + C$

13. $\displaystyle\int \csc x \cot x\,dx = -\csc x + C$

However, using the basic indefinite integrals, we are not able to evaluate an integral such as

$$\int \cos(2x + 3)\,dx$$

In order to find this integral, we use a method called **U-Substitution Rule**, which is a valuable tool to find the antiderivative of functions resulted from the Chain rule.

U-Substitution Rule

Suppose f is continuous on $[a, b]$ and $u = g(x)$ is a differentiable function on $[a, b]$. Then, $du = g'(x)dx$. Thus,

$$\int f\big(g(x)\big) g'(x)\, dx = \int f(u)\, du$$

 Tip

1. The most important part of the U-Substitution Rule is to change from the variable x to a new variable u, and change from dx to du. The new integral $\int f(u)\, du$ after the U-Substitution becomes one of the basic indefinite integrals so that you can evaluate the new integral at ease.

2. du is the differential. For instance, if $u = x^2$, then $du = 2x\, dx$.

Example 8   Evaluating an indefinite integral using the U-Substitution rule

Evaluate $\int \cos(2x + 3)\, dx$.

Solution   Let $u = 2x + 3$. Then $du = 2\, dx$ or $\dfrac{1}{2}\, du = dx$. Thus,

$$
\begin{aligned}
\int \cos(2x + 3)\, dx &= \int \cos u \frac{1}{2}\, du \\
&= \frac{1}{2} \int \cos u\, du \qquad && \int \cos u\, du \text{ is a basic indefinite integral} \\
&= \frac{1}{2} \sin u + C \qquad && \text{Substitute } u \text{ for } 2x + 3 \\
&= \frac{1}{2} \sin(2x + 3) + C
\end{aligned}
$$

Example 9   Evaluating an indefinite integral using the U-Substitution rule

Evaluate $\displaystyle\int \frac{\ln x}{x}\, dx.$

Solution   Let $u = \ln x$. Then $du = \dfrac{1}{x}\, dx$. Thus,

$$\int \frac{\ln x}{x}\, dx = \int \ln x\, \frac{1}{x}\, dx$$

$$= \int u\, du \qquad\qquad \int u\, du \text{ is a basic indefinite integral}$$

$$= \frac{1}{2}u^2 + C \qquad\qquad \text{Substitute } u \text{ for } \ln x$$

$$= \frac{1}{2}\ln^2 x + C$$

Example 10   Evaluating an indefinite integral using the U-Substitution rule

Evaluate $\displaystyle\int \frac{3}{5x - 4}\, dx.$

Solution   Let $u = 5x - 4$. Then $du = 5\, dx$ or $\dfrac{1}{5}\, du = dx$. Thus,

$$\int \frac{3}{5x - 4}\, dx = 3\int \frac{1}{5x - 4}\, dx$$

$$= 3\int \frac{1}{u}\, \frac{1}{5}\, du$$

$$= \frac{3}{5}\int \frac{1}{u}\, du \qquad\qquad \int \frac{1}{u}\, du \text{ is a basic indefinite integral}$$

$$= \frac{3}{5}\ln|u| + C \qquad\qquad \text{Substitute } u \text{ for } 5x - 4$$

$$= \frac{3}{5}\ln|5x - 4| + C$$

General Forms of the Indefinite Integrals

$$\int (ax+b)^n \, dx = \frac{1}{a(n+1)} (ax+b)^{n+1} + C, \qquad a \neq 0 \text{ and } n \neq -1$$

$$\int \cos(ax+b) \, dx = \frac{1}{a} \sin(ax+b) + C$$

$$\int \sin(ax+b) \, dx = -\frac{1}{a} \cos(ax+b) + C$$

$$\int \sec^2(ax+b) \, dx = \frac{1}{a} \tan(ax+b) + C$$

$$\int e^{ax+b} \, dx = \frac{1}{a} e^{ax+b} + C$$

$$\int \frac{1}{ax+b} \, dx = \frac{1}{a} \ln|ax+b| + C$$

15.3 Definite integrals

Definition of a Definite Integral

$\displaystyle\int_a^b f(x)\,dx$ is a definite integral of f from a to b. In the notation of $\int_a^b f(x)\,dx$, the symbol $\int dx$ is called an **integral sign**, $f(x)$ is called the **integrand**, a is called **lower limit** and b is called **upper limit**.

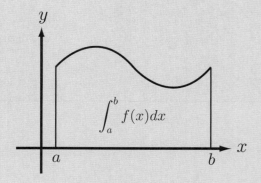

In general, $\displaystyle\int_a^b f(x)\,dx$ represents the area under curve $y = f(x)$ from a to b as shown in figure above.

Tip   Note that the area is positive if the function lies above the x-axis. Whereas, the area is negative if the function lies below the x-axis. Thus, the A_1 is positive and A_2 is negative as shown in the figure below.

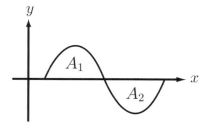

The Fundamental Theorem of Calculus

If f is continuous on $[a, b]$, then

$$\int_a^b f(x)\, dx = F(x) \Big]_a^b = F(b) - F(a)$$

Where F is any antiderivative of f such that $F' = f$.

 Tip   The Fundamental Theorem of Calculus enables us to simplify the computation of definite integrals using any antiderivative of f.

Example 11   Evaluate the integral using the Fundamental Theorem of Calculus

Evaluate $\displaystyle\int_2^1 \frac{1}{x^2}\, dx$.

Solution   Since the antiderivative of $\dfrac{1}{x^2}$ is $-\dfrac{1}{x}$,

$$\int_2^1 \frac{1}{x^2}\, dx = -\int_1^2 \frac{1}{x^2}\, dx$$

$$= \frac{1}{x}\Big]_1^2 = \left(\frac{1}{2} - \frac{1}{1}\right) = -\frac{1}{2}$$

U-Substitution for Definite Integrals

Suppose $u = g(x)$ and $du = g'(x)dx$. Then the new lower limit and the upper limit for the integration are $g(a)$ and $g(b)$, respectively. Thus, the definite integral by U-Substitution rule is given by

$$\int_a^b f\big(g(x)\big) g'(x)\, dx = \int_{g(a)}^{g(b)} f(u)\, du$$

Example 12   **Evaluating an definite integral using the U-Substitution rule**

Evaluate $\displaystyle\int_1^5 \sqrt{x-1}\, dx$.

Solution   Let $u = x - 1$. Then $du = dx$. Let's find the new lower limit and upper limit for the integration.

When $x = 1$, $u = x - 1 = 0$,   When $x = 5$, $u = x - 1 = 4$

Thus, the new lower limit and upper limit for the integration are 0 and 4, respectively.

$$\int_1^5 \sqrt{x-1}\, dx = \int_0^4 \sqrt{u}\, du = \frac{2}{3} u^{\frac{3}{2}} \bigg]_0^4 = \frac{16}{3}$$

15.4 Area between two curves

Area Between Curves

Consider the region S enclosed by two curves f and g, and two vertical lines $x = a$ and $x = b$, where f and g are continuous functions in $[a, b]$ and $f(x) \geq g(x)$ for all x in $[a, b]$ as shown in Figure 1. The area of the ith rectangle with width Δx and height $f(x_i^*) - g(x_i^*)$ is $[f(x_i^*) - g(x_i^*)]\Delta x$.

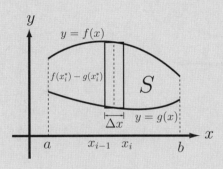

Figure 1

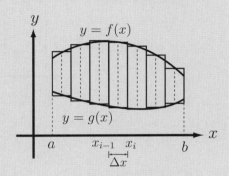

Figure 2

and the area of S can be approximated by the sum of areas of many rectangles. As $n \to \infty$, there are infinitely many rectangles as shown in Figure 2. The sum of areas of infinitely many rectangles can be written as

$$\text{Area of } S = \lim_{n \to \infty} \sum_{i=1}^{n} [f(x_i^*) - g(x_i^*)]\Delta x$$

and it can be expressed

$$\text{Area of } S = \lim_{n \to \infty} \sum_{i=1}^{n} [f(x_i^*) - g(x_i^*)]\Delta x = \int_a^b \left[f(x) - g(x) \right] dx$$

as a definite integral shown in Figure 3.

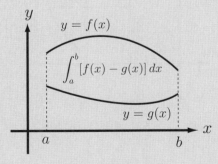

Figure 3

Finding Area between curves using Vertical Rectangles

Suppose $y = f_T$ and $y = f_B$ are continuous functions in $[a, b]$ and $f_T(x) \geq f_B(x)$ for all x in $[a, b]$ as shown below. f_T and f_B represent the top curve and bottom curve, respectively.

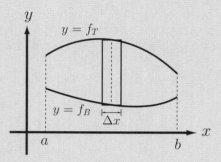

The area A enclosed by region enclosed by two curves f_T and f_B, and two vertical lines $x = a$ and $x = b$ is

$$A = \int_a^b \left[f_T - f_B \right] dx$$

Finding Area between curves using Horizontal Rectangles

Suppose $x = f_R$ and $x = f_L$ are continuous functions and $f_R \geq f_L$ for $c \leq y \leq d$ as shown below. f_R and f_L represent the right curve and left curve, respectively.

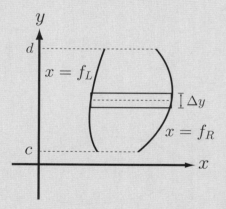

The area A enclosed by region enclosed by two curves f_R and f_L, and two horizontal lines $y = c$ and $y = d$ is

$$A = \int_c^d \left[f_R - f_L \right] dy$$

Example 13   Finding the area between curves

Find the area of the region enclosed by $y = x^2 - 3x$ and $y = 2x$.

Solution   Sketch the graphs of $y = 2x$ and $y = x^2 - 3x$ as shown below.

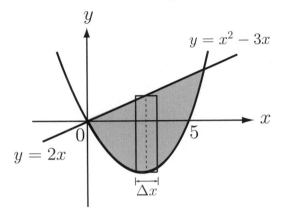

Set $x^2 - 3x = 2x$ and and solve for x to find the intersection points. This gives $x^2 - 5x = 0$ or $x(x - 5) = 0$. Thus, $x = 0$ or $x = 5$. Draw ith vertical rectangle to determine the top curve f_T and the bottom curve f_B. So, $f_T = 2x$ and $f_B = x^2 - 3x$. The total area of region enclosed by $y = x^2 - 3x$ and $y = 2x$ from $x = 0$ and $x = 5$ is

$$
\begin{aligned}
A &= \int_a^b \left[f_T - f_B \right] dx \\
&= \int_0^5 \left[2x - (x^2 - 3x) \right] dx \\
&= \int_0^5 (-x^2 + 5x)\, dx \\
&= -\frac{1}{3}x^3 + \frac{5}{2}x^2 \Big]_0^5 \\
&= -\frac{1}{3}5^3 + \frac{5}{2}5^2 \\
&= \frac{125}{6}
\end{aligned}
$$

15.5 Kinematics

Applications of Differentiation in Kinematics

If $s(t)$ is the **displacement** or position function of a particle, the first derivative of $s(t)$ represents **velocity** and is denoted by $v(t) = s'(t)$. The second derivative of $s(t)$ represents **acceleration** and is the derivative of the velocity. The acceleration is denoted by $a(t) = v'(t) = s''(t)$.

Signs for displacement s

$s < 0$	$s = 0$	$s > 0$
P is to the left of O	P is at O	P is to the right of O

where P represent a particle or an object.

Signs for velocity v

$v < 0$	$v = 0$	$v > 0$
P is moving to the left	P is instantaneously at rest	P is moving to the right

Signs for acceleration a

$a < 0$	$a = 0$	$a > 0$
velocity is decreasing	velocity = max, min, or constant	velocity is increasing

The following guidelines will help you solve a problem regarding velocity and acceleration.

1. Time at which a particle is at rest: Let $v(t) = 0$ and solve for t.

2. Time at which a particle about to change its direction: Let $v(t) = 0$ and solve for t.

3. When the particle speed up or slow down:

   - The particle speed up: when the velocity and acceleration have the same sign.
   - The particle slow down: when the velocity and acceleration have the opposite signs.

Example 14   Finding the velocity and acceleration

If the displacement function $s(t)$ is defined by $s(t) = t^3 - 3t^2 - 9t$, where t measured in seconds and s in meters.

(a) Find the velocity at $t = 3$.

(b) Find the acceleration at $t = 2$

Solution

(a) Since the velocity function is the first derivative of displacement function,

$$v(t) = s'(t) = 3t^2 - 6t - 9$$

Thus, the velocity at $t = 3$ is $v(3) = 3(3)^2 - 6(3) - 9 = 0$, which indicates that the particle is at rest or is about to change its direction.

(b) Since the acceleration function is the first derivative of velocity function,

$$a(t) = v'(t) = 6t - 6$$

Thus, the acceleration at $t = 2$ is $a(2) = 6 \ m/s^2$.

Applications of Integration in Kinematics

If $s(t)$ is the **displacement** or position function of a particle, the first derivative of $s(t)$ is **velocity** and is denoted by $v(t) = s'(t)$. Conversely,

$$\int v(t)\, dt = s(t) + C$$

The first derivative of $v(t)$ is **acceleration** and is denoted by $a(t) = v'(t)$. conversely,

$$\int a(t)\, dt = v(t) + C$$

Difference between Displacement and Total Distance

If an object moves along a straight line with position function $s(t)$, then its velocity is $v(t) = s'(t)$. Displacement and total distance are given by

$$\text{Displacement} = \int_{t_1}^{t_2} v(t)\, dt, \qquad \text{Total Distance} = \int_{t_1}^{t_2} |v(t)|\, dt$$

From the figure below, notice that A_2 is negative since the velocity function $v(t)$ lies below the x-axis. Thus, both displacement and total distance can be interpreted in terms of areas under a velocity curve.

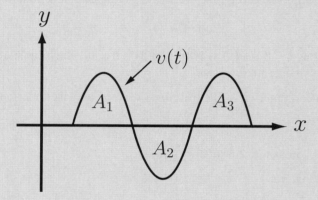

$$\text{Displacement} = \int_{t_1}^{t_2} v(t)\, dt = A_1 + A_2 + A_3, \qquad \text{Total Distance} = \int_{t_1}^{t_2} |v(t)|\, dt = A_1 + |A_2| + A_3$$

Example 15   Finding velocity and displacement

A particle moving in a straight line passes through a fixed point O with velocity 4 m s^{-1} when $t = 0$. The acceleration of the particle is given by $a(t) = 4t - 2$.

(a) Find the velocity of the particle when $t = 3$.

(b) Find the displacement of the particle from O when $t = 3$

Solution

(a) $a(t) = 4t - 2$.

$$v(t) = \int a(t) \, dt =$$
$$\int 4t - 2 \, dt$$
$$= 2t^2 - 2t + C$$

Using $v = 4$ when $t = 0$ gives $C = 4$. Thus, the velocity of the particle at time t is $v(t) = 2t^2 - 2t + 4$. Therefore, the velocity of the particle when $t = 3$ is $v(3) = 2(3)^2 - 2(3) + 4 = 16 \text{ m s}^{-1}$.

(b) The displacement of the particle from O when $t = 0$ is $s(0) = 0$. $v(t) = 2t^2 - 2t + 4$.

$$s(t) = \int v(t) \, dt$$
$$= \int 2t^2 - 2t + 4 \, dt$$
$$= \frac{2}{3}t^3 - t^2 + 4t + C$$

Using $s = 0$ when $t = 0$ gives $C = 0$. Thus, the displacement of the particle at time t is $s(t) = \frac{2}{3}t^3 - t^2 + 4t$. Therefore, the displacement of the particle from O when $t = 3$ is $s(3) = \frac{2}{3}(3)^3 - (3)^2 + 4(3) = 21 \text{ m}$.

EXERCISES

1. Find the following:

   (a) $\int \left(\sin x + \cos x \right) dx$

   (b) $\int \left(x^3 - 2x^2 + 3x - 4 \right) dx$

   (c) $\int \left(1 - \dfrac{3}{x} \right)^2 dx$

   (d) $\int \left(x + \dfrac{2}{x} + \dfrac{3}{\sqrt{x}} \right) dx$

   (e) $\int \dfrac{e^{2x} + e^x}{e^x} dx$

   (f) $\int \dfrac{1 - \cos^2 x}{(\sin x \cos x)^2} dx$

2. Evaluate the following:

(a) $\displaystyle\int_1^2 \frac{1}{2}x^4\, dx$

(b) $\displaystyle\int_1^3 \frac{1}{x^2}\, dx$

(c) $\displaystyle\int_0^1 \left(x + \sqrt{x}\right) dx$

(d) $\displaystyle\int_1^4 \left(2\sqrt{x} - \frac{1}{\sqrt{x}}\right) dx$

(e) $\displaystyle\int_0^2 e^x\, dx$

(f) $\displaystyle\int_1^2 \left(\frac{3}{x} + 2\right) dx$

3. Find the following:

(a) $\displaystyle\int (3x + 4)^7 \, dx$

(b) $\displaystyle\int 2\cos(2 - x) \, dx$

(c) $\displaystyle\int -\frac{1}{2} e^{-2x+5} \, dx$

(d) $\displaystyle\int \frac{2}{3} \sec^2(5x - 2) \, dx$

(e) $\displaystyle\int -\sin(10 - 4x) \, dx$

(f) $\displaystyle\int \frac{2}{-7x - 6} \, dx$

4. Evaluate the following:

(a) $\displaystyle\int_0^1 (2x+1)^3\,dx$

(b) $\displaystyle\int_0^7 \sqrt{3x+4}\,dx$

(c) $\displaystyle\int_1^2 \frac{6}{(5+4x)^2}\,dx$

(d) $\displaystyle\int_0^{\frac{\pi}{3}} \sin\left(2x-\frac{\pi}{6}\right)dx$

(e) $\displaystyle\int_2^5 \left(\frac{2}{x+1}-\frac{3}{1-2x}\right)dx$

(f) $\displaystyle\int_0^2 e^{\frac{1}{2}x-1}\,dx$

5. Answer the following:

   (a) Given that $y = \dfrac{x-6}{\sqrt{3x+1}}$, show that $\dfrac{dy}{dx} = \dfrac{3x+20}{2\sqrt{(3x+1)^3}}$.

   (b) Hence find $\displaystyle\int \dfrac{3x+20}{\sqrt{(3x+1)^3}}\, dx$.

6. Answer the following:

    (a) Show that $\dfrac{d}{dx}\left(\dfrac{x}{\sin x}\right) = \dfrac{\sin x - x\cos x}{\sin^2 x}$.

    (b) Hence find $\displaystyle\int_{\frac{\pi}{4}}^{\frac{\pi}{2}} \dfrac{\sin x - x\cos x}{3\sin^2 x}\, dx$.

7. Sketch the region enclosed by the following curves, and find the area of the region.

    (a) $y = x^2 + 1$, $y = x - 1$, $x = -1$, $x = 3$

    (b) $y = \cos x$, $y = \sin x$, $x = \dfrac{\pi}{4}$, $x = \dfrac{\pi}{2}$

    (c) $y = x^2$, $y = \sqrt{x}$

    (d) $x = y^2 + 1$, $x = -y^2 - 1$, $y = -1$, $y = 2$

8. Sketch the region enclosed by the following curves, and find the area of the region.

    (a) $y = e^x$, $y = \sin 2x$, $x = 0$, $x = \dfrac{\pi}{2}$

    (b) $y = (x - 3)^2$, $y = x - 3$

    (c) $y = \sqrt{x}$, $y = 2 - x$, and the x-axis.

9. The tangent to the curve $y = 4x - x^2$ at the point $(1, 3)$ intersects the x-axis at the point A.

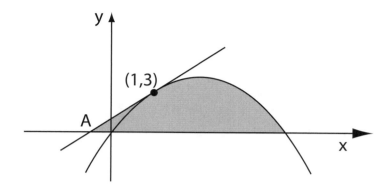

(a) Find the coordinates of A.

(b) Find the area of the shaded region.

10. A particle moves along the x-axis. The position of particle at time t given by

$$s(t) = \frac{1}{3}t^3 - 2t^2 + 3t, \qquad 0 \le t < 5$$

where t is measured in seconds and s in feet.

   (a) Find the acceleration at $t = 4$.

   (b) Find the time at which the particle changes its direction.

   (c) When is the particle speeding up?

11. Suppose a particle moves along a straight line with its velocity at time t is given by $v(t) = t^2 - 3t + 2$, where $0 \le t \le 4$ (measured in meters per second).

   (a) Find the displacement of the particle during the time period.

   (b) Find the total distance traveled by the particle during the time period. (Use a calculator)

12. A particle moving in a straight line passes through a fixed point O with velocity 12 $\mathrm{m\,s}^{-1}$ when $t = 0$. The acceleration of the particle is given by $a(t) = 4 - 2t$ for $t \geq 0$.

   (a) Find the time t at which the particle is instantaneously at rest for $t \geq 0$.

   (b) Sketch the velocity-time graph for the motion of the particle for $t \geq 0$.

   (c) Find the distance (or displacement) that the particle travelled in the first 8 seconds of its motion.

   (d) Find the total distance that the particle travelled in the first 8 seconds of its motion. (Use a calculator)

The Complete Review Book For The
IGCSE Additional Mathematics

발 행 2023년 7월 5일 초판 1쇄
저 자 이연욱
발행인 최영민
발행처 헤르몬하우스
주 소 경기도 파주시 신촌로 16
전 화 031 – 8071 – 0088
팩 스 031 – 942 – 8688
전자우편 hermonh@naver.com
출판등록 2015년 3월 27일
등록번호 제406 – 2015 – 31호

© 이연욱 2023, Printed in Korea.

ISBN   979 – 11 – 92520 – 52 – 0   (53410)